Physicochemical Characteristics of
Oligonucleotides and Polynucleotides

Physicochemical Characteristics of Oligonucleotides and Polynucleotides

BOREK JANIK

Molecular Biology Department
Miles Laboratories, Inc.
Elkhart, Indiana

IFI/PLENUM • NEW YORK–WASHINGTON–LONDON • 1971

Library of Congress Catalog Card Number 70-165692

ISBN-13: 978-1-4613-4614-2   e-ISBN-13: 978-1-4613-4612-8

DOI: 10.1007/ 978-1-4613-4612-8

© 1971 IFI/Plenum Data Corporation
Softcover reprint of the hardcover 1st edition 1971
A Subsidiary of Plenum Publishing Corporation
227 West 17th Street, New York, N. Y. 10011

United Kingdom edition published by Plenum Press, London
A Division of Plenum Publishing Company, Ltd.
Davis House (4th Floor), 8 Scrubs Lane, Harlesden, NW10 6SE, England

PREFACE

   Physicochemical studies on polynucleotides and their components
are a relatively new field which, almost daily, attracts an ever-
increasing number of scientists. To date, however, only a limited
effort has been made to compile the vast amount of physicochemical
data available into a useful format. I originally undertook this
compilation of data to supplement my own research efforts. But it
soon became evident that my labors might be of benefit to other
workers in this complex area, and this volume is the result.

   The information cited in this manuscript covers the literature
up to the end of 1970. The present compilation cannot possibly
list all recorded dissociation constants, extinction coefficients,
and T_m values of polynucleotides, oligonucleotides, and their complexes. As outlined in the "Explanation of Tables," some discre-
tion had to be exercised in the selection of oligonucleotides and
polynucleotides for inclusion. Every effort has been made, however,
to provide more than just a representative cross section of the
available literature.

   I should like to take this opportunity to call the reader's
attention to other compilations of selected physicochemical data on
polynucleotides and to some relevant reviews on this subject. T_m
values of polynucleotides are listed in a review on physical and
chemical properties of nucleic acids by Felsenfeld and Miles (56)
(literature covered to August, 1966). The "Handbook of Biochemistry," edited by H. A. Sober (235), covers the literature up to
1967 and includes also tables on the spectral characteristics and
dissociation constants of bases, nucleosides, and nucleotides,
together with optical rotatory parameters of ribonucleotides and
oligoribonucleotides. Tables of absorbance ratios, hyperchromicities, and maximum and minimum wavelengths of nucleotides and
oligonucleotides may be found in two easily accessible sources
(235, 264). Therefore, these data are not included here. Absorption spectra of nucleotides of minor bases and selected dinucleotides are listed in Venkstern's book (276). The reader is referred
to specialized reviews (56, 173, 212, 268, 283a) and a book (242)

for the physical and chemical properties and characterization of
polynucleotides and oligonucleotides.

The author wishes to thank the management of Miles
Laboratories, Inc., for generous support; without their aid this
book would not have been possible. Special mention is due Dr.
L. A. Underkofler, Director of the Molecular Biology Department
of Miles Laboratories, for his understanding, support, and stimulus
for writing this book. I also wish to thank my colleagues and
friends for their stimulating discussions, criticism, and help in
sorting through the intricacies of this subject. I am especially
grateful to Dr. Michael P. Kotick, Dr. Thomas H. Kreiser, Mrs.
Gail F. Martin, and Mr. Robert W. Martin. I am greatly indebted
to Mrs. June L. Rieth for secretarial assistance, and to numerous
typists, particularly Mrs. Gloria Vittner and Miss Linda Pavoni,
for their skill and patience in typing the difficult text from
nearly-illegible manuscript. Finally, I would like to thank my
wife, Alice, for her patient understanding and endurance.

The author would be grateful to users of this book for any
comments and suggestions.

                                                             Borek Janik

CONTENTS

EXPLANATION OF TABLES

Content of Tables

The tables consist of three major sections, presenting data on dissociation, spectral, and melting properties, respectively, of polynucleotides, oligonucleotides, and their complexes. The tables encompass polynucleotides which are of synthetic origin, i.e., those polynucleotides which do not fall in the class of natural DNAs and RNAs. For comparative purposes, however, data are included on several preparations of poly d(A-T) isolated either from calf thymus or crab male sex organs (193, 194). Polynucleotides (and long-chain oligonucleotides) of the type synthesized by the Khorana group (cf. reference 75) are not included because they more closely resemble the natural RNAs and DNAs. Chemically modified polynucleotides are listed either as copolymers when the extent of modification was reported in the literature, or as homopolymers, when they were claimed as 100% converted or when data on the extent of modification could not be located in the cited reference. Data on synthetic polynucleotides from early papers in this field are not included wherever the polymeric nature of the so-called polynucleotides is questionable; such polynucleotides were often oligomers of unspecified chain length.

Oligonucleotides included in the tables were either obtained by synthetic procedures or by partial hydrolysis of nucleic acids or synthetic polynucleotides, or isolated directly from natural sources. Data are not included on oligonucleotides whose identity was not fully confirmed (e.g., reference 267), nor on derivatives which are intermediates in oligonucleotide syntheses, for example, those containing various N- and O- protecting groups.

Inclusion of data on complexes of poly- and oligonucleotides with various organic and inorganic substances, e.g., dyes, hormones, antibiotics, proteins, peptides, heavy metals, etc., is mostly limited to representative examples. Calculated data are generally not included (e.g., reference 31) unless they are accompanied by experimental data (e.g., reference 28).

1

Every effort has been made to provide all relevant data on experimental conditions, methods, etc., and at the same time to keep the tables to reasonable size. In order to limit size, the inclusion of the cited authors' interpretations of experimental data is limited to cases when explicit and relevant information is given in the reference; e.g., the melting of a polynucleotide is designated as cooperative only when it is so stated in the reference. No attempt has been made to draw conclusions from the compiled data.

The first column in each table gives symbols of substances or their complexes; the formation of symbols and the order in which entries are arranged are described in the following two subsections. The formation and content of other columns are in accord with established practice; peculiarities are explained in prefaces to individual sections. Columns which precede those showing the dissociation, spectral, or melting characteristics provide information on methods and/or experimental conditions. Any additional information is given under Notes or Footnotes.

Nomenclature and Symbols

Abbreviations and symbols for common polynucleotides and their constituents are those recommended by the IUPAC--IUB Commission on Biochemical Nomenclature in "Abbreviations and Symbols for Nucleic Acids, Polynucleotides, and Their Constituents. Recommendations 1970" (116).

Monomeric Forms. The common monomeric forms (e.g., as they are shown in complexes of bases, nucleosides, and nucleotides with polynucleotides) and base moieties in oligo- and polynucleotide analogues which do not contain nucleotides as monomeric units (e.g., polyvinylanalogues) are designated by the following symbols (three-letter symbols are used for purine and pyrimidine derivatives):

Acrid	acridine
Ade, Ado	adenine, adenosine
Cyt, Cyd	cytosine, cytidine
Gua, Guo	guanine, guanosine
Hyp, Ino	hypoxanthine, inosine
Imid	imidazole
Nuc	a nucleoside
Pur, Puo	purine, purine nucleoside
Pyr, Pyd	pyrimidine, pyrimidine nucleoside
Pyrid	pyridine
Thy, Thd	thymine, ribosylthymine
Ura, Urd	uracil, uridine
Xan, Xao	xanthine, xanthosine

Nucleotides are expressed as phosphoric esters, such as Ado-3'-P for adenosine-3'-monophosphate. Symbols for monomers containing substituted purine and pyrimidine bases and sugar residues other than ribose are formed in accord with rules subsequently described for one-letter symbols.

Oligonucleotides are represented by a combination of one-letter symbols for phosphate and nucleoside residues. A monosubstituted (terminal) phosphoric residue is represented by a small p. A diesterified phosphate residue is represented by a hyphen when the sequence is known and by a comma when the sequence is unknown; it is considered to be attached to the 3' oxygen atom on the left and to the 5' oxygen atom on the right. Thus, a hyphen represents a 3' → 5' linkage. Analogously, p to the left and to the right of the nucleoside symbol represents a 5' and 3' terminal phosphate residue, respectively. A 2':3'- cyclic phosphate residue is indicated by >p. For other types of linkages the single phosphate residue symbol is combined with locants, e.g., A2'-5'A and C-G2'p. These locants precede a chain if the internucleotide linkage is identical throughout.

Common ribonucleoside residues are designated by single capital letters (one-letter symbols), as follows:

A   adenosine
C   cytidine
G   guanosine
I   inosine
N   unspecified or any nucleoside
Q   pseudouridine (5-ribosyluracil)
T   ribosylthymine
U   uridine
X   xanthosine

Nucleoside residues containing uncommon purine and pyrimidine derivatives are symbolized, wherever possible and practical, as derivatives of common nucleosides, e.g., 2,6-diaminopurine ribonucleoside is abbreviated as 2-aminoadenosine, n^2A. These symbols, as well as those for substituents (the latter are subsequently listed), were suggested in accord with the principles set out by the above-mentioned commission. Some symbols for nucleoside residues (also subsequently listed) were formed to preserve the common name of the nucleoside and may deviate from the above rules.

Sugar residues other than ribose are represented by a prefix, e.g., d and a for 2'-deoxyribosyl and arabinosyl residues, respectively. Thus: aC-dG, arabinocytidylyl-deoxyguanosine; d(A-T), deoxyadenylyl-thymidine; d(A2'-5'C-G-Cp), a tetranucleotide with all deoxy residues, a single 2'-5' linkage (between A and C) and

terminal 3' phosphate on C. The L enantiomers of the common
D-sugars are designated by a subscript L preceding the symbol for
the nucleoside residue (e.g., $_L A$ - $_L A$, L-adenylyl-3'-5'-L-adenosine).

Polynucleotides. The common polynucleotides are represented
by "poly" which precedes the repeating unit symbolized by one-letter
symbols for nucleosides; thus poly A, poly d(A-T), and poly (A,C)
are for polyadenylate, alternating copolymer of dA and dT, and
random copolymer of A and C, respectively.

Oligo- and Polynucleotide Analogues. Polynucleotides containing a phosphorothioate backbone were symbolized according to
recommendations by Dr. W. E. Cohn (Director of the National
Academy of Sciences--National Research Council, Office of Biochemical Nomenclature, Oak Ridge National Laboratory, Oak Ridge,
Tenn.) as these were stated in the literature (52). The phospho-
rothioate linkage $-O-\overset{\overset{S}{\|}}{\underset{\underset{O_-}{|}}{P}}-O-$ is symbolized by $_s$ (the hyphen represents
a phosphodiester 3'-5' linkage and subscript s represents the thio
group); thus poly ($_s$U) is a homopolymer containing uridine-5'-O-
thiophosphate residues, and poly ($_s$C-I) is an alternating copolymer
containing cytidine-5'-O-thiophosphate and inosine-5'-phosphate
residues.

Dinucleotide analogues in which the bases are connected by a
polymethylene chain are symbolized by B-C_n-B', where B and B' are
9-substituted adenine or guanine or 1-substituted cytosine, thymine,
or uracil residues and n is the chain length in -CH_2- units (e.g.,
Ade-C_3-Cyt).

Polyvinyl analogues are symbolized similarly to common polynucleotides. In order to express their genesis the symbol for the
repeating unit is that for the monomeric form (e.g., v^9Ade, 9-
vinyladenine, or Ade-$CH=CH_2$) rather than that for the actual
repeating unit (e.g., (adenine-9-yl) ethylene or Ade-$CH-CH_2$-).

Symbols for some uncommon oligonucleotides which do not fit
into any of the above categories are listed below.

Association (noncovalent) between two or more polynucleotide
chains or their constituents is indicated by the center dot (e.g.,
poly A·2 poly U); absence of definite information on association
is indicated by the comma (e.g., C(pA)$_5$, G(pU)$_5$); absence of
association is indicated by the plus sign (e.g., poly dC + poly
d(C,I)).

Symbols for Substituents

ac	acetyl	io	iodo
acn	acetylamino	m	methyl
br	bromo	mo	methoxy
c	deaza	ms	methylthio
cl	chloro	n	amino
e	ethyl	s	thio or mercapto (sU =
fl	fluoro		thiouridine)
h	dihydro (hU = di-	sba	3,6-disulfobenzenazo
	hydrouridine)	sca	semicarbazido (NH$_2$NH·CO·NH-
he	hydroxyethyl		replaces NH$_2$-; e.g., ssa^{4}C =
hm	hydroxymethyl		4-semicarbazido-2-ketopyrimi-
hn	hydroxyamino		dine)
ho	hydroxy	v	vinyl
ib	isobutyl		

Locants are indicated by superscript, e.g., m^{7}G indicates a 7-methylguanosine residue; and multipliers simply by giving the proper number of locants (this is a deviation from IUPAC--IUB rules), e.g., m^{2,6}A indicates a 2,N^6-dimethyladenosine residue. Symbols of substituents modifying the sugar residue are placed immediately to the right of the nucleoside symbol and indicate substitution at the 2' position unless otherwise specified.

Uncommon Symbols for Some Polynucleotides and Their Components

Nucleoside residues:

c^{7}A	7-deazaadenosine, tubercidin
F	formycin, 7-amino-3(β-D-ribofuranosyl)-pyrazolo-(4,3-d) pyrimidine
isoA	isoadenosine, 3-ribosyladenosine

Oligonucleotides:

a(s$^{2',8}$A-s$^{2',8}$A)	8,2'-anhydro-8-mercapto-9-β-D-arabinofuranosyladenylyl-(3'-5')-8,2'-anhydro-8-mercapto-9-β-D-arabinofuranosyladenine
A-U4ss4U5'-3'A	bis[adenylyl-(3'-5')-4-deoxy-4-uridinyl] disulfide
G-U4ss4U5'-3'G	bis[guanylyl-(3'-5')-4-deoxy-4-uridinyl] disulfide
U4ss4U	bis(4-deoxy-4-uridinyl) disulfide
d(A-T)$_n$	circular alternating oligomer of dA and dT

$d^{3'}[(m^{6,6}An^{3'})-$ [6-dimethylamino-9-β-D-(3'-amino-3'-deoxy)-
   $(m^{6,6}Aacn^{3'})]$ ribofuranosylpuryrinyl]-5'P5'-[6-dimethyl-
                               amino-9-β-D-(3'-acetamido-3'-deoxy)-ribo-
                               furanosyl purine]

Polynucleotides:

   poly A(CH$_2$0)   formaldehyde treated poly A
   poly A(ox)   poly A oxidized at N-1 by monoperphthalate
   poly C(ox)   poly C oxidized at N-3 by monoperphthalate

How to Use the Tables

    Entries are arranged in each section based on the following
criteria (listed in an order of decreasing inclusiveness):

    a) The common monomeric forms (nucleoside residues in most
cases) of parent oligomers and polymers are in alphabetical order,
i.e., A, C, G, I, T, U, and X. Derivatives and analogues are listed
following the parent compound; isoadenine (isoA) and formycin (F),
and pseudouridine (Q) are considered as analogues of adenosine and
uridine, respectively. Monomers are arranged in alphabetical order,
wherever possible, in symbols of compounds containing more than one
kind of monomer, e.g., poly(C-G) and poly C·poly I are used instead
of generally accepted poly (G-C) and poly I·poly C, respectively.
Such compounds are then arranged alphabetically by the monomer symbol. Complexes of polynucleotides with their constituents are also
listed alphabetically by the polymeric component and then by the
constituents. [Exception: Grouping of oligomers in Section 2.1.2
(Spectral Data of Heterooligomers) is by dimers, trimers, and tetra-
and higher oligomers rather than by alphabetical order].

    b) Order of increasing chain length (oligomers).

    c) 2'-Deoxyribosyl compounds are listed after ribosyl com-
pounds followed by derivatives including other sugar moieties, e.g.,
arabinosyl.

    d) Derivatives of parent compounds
        1. substitution on the heterocyclic moiety;
        2. substitution on the sugar moiety.

    e) Type of phosphate linkage and presence of end phosphate
groups (oligonucleotides) in order, from more common and less
complex to less common and more complex situations.

    f) Order of increasing pH.

g) Order of increasing ionic strength, medium components concentration and complexity of medium composition.

The above order of criteria should serve as a general guideline; it is changed wherever aspects followed in the cited reference call for such a change.

Abbreviations

Numbers in parentheses denote the section in whose preface a more detailed definition is given. Abbreviations from the Method columns are listed in the preface to each section.

Ac	acidic (in the pH column)
acet.	acetate buffer
a.h.	alkaline hydrolysis (2)
ac.h.	acid hydrolysis (2)
B	E_λ is as per 1 mol base per liter
bor.	borate buffer
b.t.	broad thermal transition (3)
buff.	buffer
c.	coil
cacd.	cacodylate buffer
carb.	carbonate buffer
citr.	citrate buffer
co.	cooperative transition
concn.	concentration
cx.	complex
diox.	dioxane
DMSO	dimethylsulfoxide
d.s.cx.	double-stranded complex
d.s.hx.	double-stranded helix
$E(\lambda)$	extinction coefficient (molar absorbancy) x 10^{-3} at wavelength λ
$E_{min}(\lambda)$	extinction coefficient x 10^{-3} at the minimum; wavelength of the minimum (mμ) is given in parentheses
e.g.	ethylene glycol
e.h.	enzymatic hydrolysis (2)
form.	formate
gly.	glycine buffer
H	hypochromism (2)
h.p.hx.	half-protonated helix
hper.	hyperchromicity (2,3)
hper.(hx-c)	hyperchromicity of the thermally induced helix-coil transition (2)
hpo.	hypochromicity (2)
hx.	helix

I	ionic strength, $I = \frac{1}{2}\Sigma C_i \cdot Z_i$, where C_i is molar concentration of the ion in solution and Z_i is its charge
incompl.	incomplete, e.g., incomplete thermal transition (3)
irev.	irreversible transition
m	chain length of oligomers in monomeric units
M	molecular weight x 10^{-5} of a polynucleotide (polynucleotide may be specified by a subscript), e.g., molecular weight of 5 x 10^5 of poly A is shown as $M_A = 5$
m_3n	trimethylamino group in polymethylene diamines, e.g., $m_3nC_znm_3$
max	absorption maximum, e.g., 260max = maximum at 260 mμ
min	absorption minimum, e.g., 230min = minimum at 230 mμ
mixt.	mixture
n	chain length of oligomers in monomeric units or amino group in polymethylene diamines, nC_zn
N	neutral (in the pH column) or any nucleotide (in the Note column)
nC_zn	polymethylene diamine; z is the number of -CH$_2$- units, e.g., nC_2n, ethylene diamine
nco.	noncooperative transition
n.p.c.	nonprotonated coil
P	E_λ is as per 1 mol phosphorus per liter
p.a.	phosphorus analysis (2)
part.	partially
p.d.h.	pancreatic DNase hydrolysis (2)
p.g.	propylene glycol
pH$_m$	transition pH (1)
phosp.	phosphate buffer
ppn.	preparation
ppt.	precipitate
p.r.h.	pancreatic RNase hydrolysis
rev.	reversible transition
rm.	room temperature
sh	shoulder, e.g., 280sh = shoulder at 280 mμ
sperd.	spermidine
sperm.	spermine
SSC	standard saline citrate (0.15 M NaCl, 0.015 M Na$_3$citrate); 1/10 SSC, 10 times diluted SSC
s.s.hx.	single-stranded helix
su.	sucrose
s.v.h.	snake venom phosphodiesterase hydrolysis
t.	thermal transition
t_1, t_2	first and second step in a multistep thermal transition

Tris.	tris (hydroxymethyl) aminomethane
t.s.cx.	triple-stranded complex
VAR	variable; VAR is used in connection with parameters describing experimental conditions that were varied to produce different values of the measured quantity and is used to denote that quantity. The measured quantity was presented in a graphical or tabular form in the cited reference.
z	chain length of polymethylene diamines in $-CH_2-$ units
$\lambda_{max}(2)$	maximum wavelength (mμ) at pH 2
$\ast^{\pi}\underline{\quad}$	these symbols are explained in the preface to Section 2
( )	values of the measured quantity given in parentheses are those which were estimated by the compiler from graphs reported in the cited reference; hence, they have to be considered as approximate
[ ]	molar concentration; usually used in connection with VAR, e.g., VAR[Na^+]

1. DISSOCIATION PROPERTIES

INTRODUCTION

pK_{mono} is the negative logarithm of the apparent dissociation constant, pK_a, of a base, B or B$^-$, in terms of the equilibrium of the base and its conjugated acid, BH^+ or BH, with a solvated proton,

$$Baq + H_3Oaq \rightleftharpoons BH^+aq + nH_2O$$
$$B^-aq + H_3O^+aq \rightleftharpoons BHaq + nH_2O$$

where the base is the constituting monomer, nucleoside-5'-mono-phosphate, unless otherwise indicated. The pK_{mono} values for nucleotides are only those for the nucleoside moiety, not for phosphate ionization.

pK_{oligo} and pK_{poly} are apparent pK_a constants of the nucleo-side moiety of oligomer and polymer, respectively, and their complexes.

Values of the transition pH, pH_m, which is defined as a mid-point of the pH-induced helix--random coil transition or complex dissociation, may also be given in the pK_{mono} and pK_{poly} columns. However, the pK values are shown as pH_m values (in the Note column) only when pertinent information is given in the cited reference. The pH-induced complex dissociation may be described in an abbre-viated form, e.g., $A \cdot U \rightarrow A \cdot 2U + A \cdot A$ is used in place of

$$2(poly \ A \cdot poly \ U) \rightarrow poly \ A \cdot 2 \ poly \ U + \tfrac{1}{2}(poly \ A \cdot poly \ A)$$

The methods of determining pK values (or parameters measured) are abbreviated as follows:

CD   circular dichroism; the wavelength (mμ) at which ellipticity was measured may be given following CD, e.g., CD275

EB   electrical birefringence

LS   light scattering

ORD   optical rotatory dispersion; the wavelength (mμ) at which rotation was measured may be given following ORD, e.g., ORD280.

P   potentiometry
S   spectrophotometry; wavelength (mμ) at which optical
        density was measured may replace S
V   viscometry

1. DISSOCIATION PROPERTIES

1.1. OLIGONUCLEOTIDES

1.1.1. Homooligomers

Oligomer	Method	T °C	Medium	pK_{mono}	pK_{oligo}	Note	Ref.
A–A	P	20	0.1 M NaCl		3.50±0.02		229
A–Ap	P	20	I = 0.1 M		6.16		113
$(Ap)_n$	260				VAR	n=1-45	171
A_2	260		0.15 M NaCl	4.13	3.34		177
A_3	260		0.15 M NaCl		3.40		177
A_5	260		0.15 M NaCl		3.40		177
A_6	260		0.15 M NaCl		2.89		177
A_7	260		0.15 M NaCl		2.23;4.12		177
A_8	260		0.15 M NaCl		2.02;4.55		177
A_{12}	260		0.15 M NaCl		3.05;5.17		177
A_{30}	260		0.15 M NaCl		3.14;5.48		177
A_{45}	260		0.15 M NaCl		3.37;5.51		177
A_{60}	260		0.15 M NaCl		5.58		177
A_{200}	260		0.15 M NaCl		5.54		177
C–C	ORD	20	I = 0.1 M		4.4		113
U–U	260	20	0.1 M NaCl		9.33±0.02		229
	S	20±0.2	I = 0.01 M	9.85	9.63	9.38[a]	42
	S	20±0.2	I = 0.03 M	9.70	9.52	9.32[a]	42

Oligomer	Method	$T \atop °C$	Medium	pK_{mono}	pK_{oligo}	Note	Ref.
U–U (Con't)	S	20±0.2	I = 0.1 M	9.515	9.40	9.275[a]	42
	S	20±0.2	I = 0.3 M	9.33	9.27	9.18[a]	42
	S	20±0.2	I = 1 M	9.12	9.15	9.11[a]	42
$(Up)_2$	260	20	0.1 M NaCl	9.42±0.02	9.55+0.02		231
$(Up)_3$	260	20	0.1 M NaCl		9.62±0.02		231
$(Up)_4$	260	20	0.1 M NaCl		9.66±0.02		231
$(Up)_5$	260	20	0.1 M NaCl		9.68±0.02		231
$(Up)_6$	260	20	0.1 M NaCl		9.70±0.02		231
$(Up)_7$	260	20	0.1 M NaCl		9.73±0.02		231
$(Up)_8$	260	20	0.1 M NaCl		9.72±0.02		231
$(Up)_9$	260	20	0.1 M NaCl		9.74±0.02		231
$(Up)_{10}$	260	20	0.1 M NaCl		9.73±0.02		231
Poly U	260	20	0.1 M NaCl		9.73±0.02		231
$a(Up)_{15-20}r(Up)$	S		0.1 M NaCl, 0.05 M Na cacd.	9.32	9.73		203

Oligomer	Method	T °C	Medium	pK$_{oligo}$	Note	Ref.
			1.1.2. Heterooligomers			
A-U	260	20	0.1 M NaCl	3.70±0.05; 9.48±0.02		229
C-U-C	S	20±0.2	I = 0.01 M	9.75		42
	S	20±0.2	I = 0.03 M	9.62		42
	S	20±0.2	I = 0.1 M	9.47		42
	S	20±0.3	I = 0.3 M	9.32		42
	S	20+0.2	I = 1 M	9.15		42
U-A	260	20	0.1 M NaCl	3.85±0.05; 9.35±0.02		229
U-C	S	20±0.2	I = 0.01 M	9.61		42
	S	20±0.2	I = 0.03 M	9.53		42
	S	20+0.2	I = 0.1 M	9.38		42
	S	20±0.2	I = 0.3 M	9.28		42
	S	20±0.2	I = 1 M	9.16		42

1.2. COMPLEXES OF POLYNUCLEOTIDES WITH THEIR COMPONENTS

Complex	Method	T °C	Medium	$pK_{compl.}$	Note	Ref.
Poly A Complexes						
Pur	ORD436	20	Na phosp.; I = 0.2 $\underline{M}$	5.30		213
Poly C Complexes						
Guo	P	12	0.01 $\underline{M}$ NaCl	(6.3)		216
2 Poly C Complexes						
Guo-3'-P	CD275	5	1 $\underline{M}$ Na$^+$	(3.7);(5.2)		216
	CD255	5	1 $\underline{M}$ Na$^+$	(5.3)		216
	275	5	1 $\underline{M}$ Na$^+$	(5.8)		216
	P	5	1 $\underline{M}$ Na$^+$	(5.6)		216
GMP	ORD265	5	1 $\underline{M}$ Na$^+$	(5.35)		216
	ORD290	5	1 $\underline{M}$ Na$^+$	(5.4)		216
	256	5	1 $\underline{M}$ Na$^+$	(5.4);(5.75)		216
	270	5	1 $\underline{M}$ Na$^+$	(5.4);(5.7)		216

1.3.　POLYNUCLEOTIDES

Polymer	Method	$T_{°C}$	Medium	pK_{mono}	pK_{poly}	Note	Ref.
			1.3.1.　Homopolymers				
Poly A		20	1 mM KCl		6.8		268 loc. cit.
	LS		0.01 M Na acet.		5.5		240
	S	20	0.01 M NaCl		6.22	A·A → A+A	41
	S	20	0.02 M NaCl		6.23	A·A → A+A	41
	S	20	0.03 M NaCl		6.22	A·A → A+A	41
	S	20	0.04 M NaCl		6.14	A·A → A+A	41
	S	20	0.05 M NaCl		6.10	A·A → A+A	41
	S	20	0.1 M NaCl		5.95	A·A → A+A	41
	S	20	0.2 M NaCl		5.84	A·A → A+A	41
	S	20	0.3 M NaCl		5.77	A·A → A+A	41
	S	20	0.45 M NaCl		5.72	A·A → A+A	41
	S	20	0.6 M NaCl		5.64	A·A → A+A	41
	S	20	1 M NaCl		5.55	A·A → A+A	41
	S		VAR[NaCl]		b		41
	P	VAR	0.001,0.01 and 0.15 M KCl, resp.		VAR		91
		20	1 mM KCl		6.8		271
		26	1 mM KCl		6.4	co.	240
	257		0.01 M NaCl, 0.02 M acet.		(6.0)		8
		26	0.1 M KCl		6.0	co.	240
	P;259	25	0.1 M KCl		(5.9)		206
	P;259	10	0.1 M KCl		(6.0)		206
	P	23±1	0.1 M NaCl		3.2; 5.8–5.9		2
	S^c	23±1	0.1 M NaCl or 0.1 M Na citr.		5.87		2

Polymer	Method	T °C	Medium	pK$_{mono}$	pK$_{poly}$	Note	Ref.
Poly A (Con't)							
	300		0.1 M NaCl		5.8	co.	125
	259	18.8	0.1 M KCl,0.01 M acet. or carb.		5.9		241
	283	rm.	0.15 M KCl	3.8±0.1	5.95	co.	12
	250		0.15 M NaCl		5.79		275
		20	0.15 M KCl		6.0		268 loc. cit.
	260		0.15 M NaCl	4.13	5.54		177
	ORD436	20	Na phosp. or Na acet., I = 0.2 M		5.65		213
	260	VAR			VAR		45
		VAR	VAR [Na$^+$]		VAR		78
	EB		Mg^{++}		5.9	Mg/A=1	117
	S				6.1	co.	46
Poly m^{1}A	S				5.1		180
Poly m^{6}A	S				~ 3.3		180
Poly he^{6}A	250			3.53	3.4		275
Poly n^{2}m^{6}A	P		0.1 M Na$^+$	4.3^d	5.3	co.	102
Poly Am	283		0.15 M KCl	4.0±0.1	6.40	co.	12
Poly Aac	300		0.1 M NaCl		(4.9)	88% acet.	125
Poly isoA	277		0.15 M NaCl	5.87^e	5.75	nco.	178
Poly dA	P	23±1	0.1 M NaCl		~ 3.2; 4.4-4.5		2
	S^c	23±1	0.1 M NaCl or 0.1 M Na citr.		4.40		2
	ORD240, 252:5,279	22	1 mM Na acet.		5.3		268
	ORD240 252,279	20	0.2 M NaClO$_4$, 0.02 M Na acet.		4.4		268

Polymer	Method	T °C	Medium	pK$_{mono}$	pK$_{poly}$	Note	Ref.
Poly C	P	12	0.01 M NaCl		(6.0)		36
	P	25	0.05 M NaCl	4.2[e]	3.2;5.8		67
	P		0.1 M Na+		5.7	co.	80
	P; 275,243	25	0.1 M KCl	4.3[e,g]	3.0;5.7	co.	80
	S	90	0.13 M Na+	4.3	4.37		77
	280		0.15 M NaCl	4.33[d]	5.82		85
	270	5	1 M Na+		(5.7)		255
	P		VAR [Na+] (0.05-1.0 M Na+)		VAR		260
Poly m^{1}C				8.8[d]	(7.5);9.4	f	266
Poly m^{4}C	282		0.1 M NaCl, 0.01 M buff.	4.25	(3.5)		22
Poly m^{4,4}C	287		0.1 M NaCl, 0.01 M buff.	4.0	(3.9)		22
Poly ho^{4}C	S			2.26[d]	~2.7		119
Poly sca^{4}C	S			2.7[d]	~3.5		119
Poly br^{5}C	310		0.03 M NaCl		4.70	co.	175
	310		0.15 M NaCl	2.61(300)[h]	4.43	co.	175
	310		0.3 M NaCl		4.31	co.	175
	S	25	0.20 M Na+		(5.2)	rev.	106
Poly io^{5}C	295		0.15 M NaCl	3.01(315)[h]	5.00	co.	175
Poly Cm	S	22		4.09			121
	S				~4.6	co.	153
Poly dC	P	25	0.05 M Na+		7.5	rev.	106
	S	25	0.05 M Na+		(7.6)	rev.	106
	P		0.05 M Na+		7.2	co.	107

Polymer	Method	T °C	Medium	pK_{mono}	pK_{poly}	Note	Ref.
Poly d(m^5C)	245, 276		0.15 $\underline{M}$ NaCl, 0.1 $\underline{M}$ phosp.		(7.3)	co.	287
Poly d($m^{4,5}$C)	245, 281		0.15 $\underline{M}$ NaCl, 0.1 $\underline{M}$ phosp.	4.0	4.4	nco.	287
Poly d(e^4m^5C)	245, 281		0.15 $\underline{M}$ NaCl, 0.1 $\underline{M}$ phosp.	4.0	4.4	nco.	287
Poly G	270		0.01 $\underline{M}$ NaCl		11.86	co.	195
	270		0.05 $\underline{M}$ NaCl		11.76	co.	195
	290		0.1 $\underline{M}$ NaCl		3.0	nco.	195
	270		0.15 $\underline{M}$ NaCl		11.43	co.	195
	270		0.25 $\underline{M}$ NaCl	9.30	11.32	co.	195
	270		0.15 $\underline{M}$ NaCl	9.30	11.43	co.	176, 180, 196
	275		0.15 $\underline{M}$ NaCl		11.5	co.	179
	270	25	0.2 $\underline{M}$ Na^+	9.4^i	11.2	co.	62
			VAR [Na^+]		VAR		174
Poly m^1G	S		0.15 $\underline{M}$ NaCl	< 3	4.25		196
Poly m^7G	258		I = 0.02 $\underline{M}$	7.1	> 9.8	30%G^q	86
	S		0.15 $\underline{M}$ NaCl	6.88^d	~10	j	180
Poly $m^{2,2}$G	280		0.15 $\underline{M}$ NaCl	9.6	10.1	co.	198
	310		0.15 $\underline{M}$ NaCl	~3.0	< 3.0	nco.	198
Poly $m^{2,2,7}$G	S			7.4^e	~7.4	nco.	198
Poly br^8G	275		1 m$\underline{M}$ NaCl		10.5	nco.	179
	275		0.15 $\underline{M}$ NaCl	9.45	10.8	nco.	179
	275		1 $\underline{M}$ NaCl		11.8	nco.	179

Polymer	Method	T °C	Medium	pK_{mono}	pK_{poly}	Note	Ref.
Poly I	248		0.15 $\underline{M}$ NaCl	8.75	10.24		176,178, 198
Poly m^7I	S	25	0.15 $\underline{M}$ NaCl	6.40[d]	6.80	j	180
Poly U	S	20	0.00 $\underline{M}$ Na$^+$		11.32		174
	S	20	0.03 $\underline{M}$ Na$^+$		10.33		174
	S	20	0.15 $\underline{M}$ Na$^+$	9.40(9.5)	9.81		157,174
	S	20	0.30 $\underline{M}$ Na$^+$		9.56		174
	S	20	1.00 $\underline{M}$ Na$^+$		9.24		174
	S	2	1.0 $\underline{M}$ Na$^+$		9.60		174
	S	20	3.00 $\underline{M}$ Na$^+$		9.02		174
	S	20	0.0 $\underline{M}$ NaCl, 2 m$\underline{M}$ Tris		10.60		230
	S	20	0.01 $\underline{M}$ NaCl, 2 m$\underline{M}$ Tris		10.33		230
	S	20	0.1 $\underline{M}$ NaCl, 2 m$\underline{M}$ Tris	9.42	9.72	k	230,231
	S	20	0.5 $\underline{M}$ NaCl, 2 m$\underline{M}$ Tris		9.41		230
	S	20	VAR [Na$^+$]		VAR		230
	S	20±0.2	I = 0.01 $\underline{M}$	10.76	9.85	9.38[d]	20
	S	20±0.2	I = 0.03 $\underline{M}$	10.42	9.70	9.32[d]	20
	S	20±0.2	I = 0.1 $\underline{M}$	10.05	9.51_5	9.27[d]	20
	S	20±0.2	I = 0.3 $\underline{M}$	9.71	9.33	9.18[d]	20
	S	20±0.2	I = 1 $\underline{M}$	9.33	9.12	9.11[d]	20
	S	20	I = 0.05 $\underline{M}$		10.14		279
	P	20	I = 0.2 $\underline{M}$		9.70		279
	S	20	I = 1.0 $\underline{M}$		9.43		279
	S	20	I = 2.0 $\underline{M}$		9.33		279
	260	20	0.1 $\underline{M}$ NaCl		9.73±0.02		48
	259	18.8	0.1 $\underline{M}$ KCl, 0.01 $\underline{M}$ acet. or carb.		9.5		241
	S		0.15 $\underline{M}$ NaCl		9.60		178

Polymer	Method	T °C	Medium	pK_{mono}	pK_{poly}	Note	Ref.
Poly U (Con't)							
	S			9.25			58
	P			9.17			140
	S	20	1 M KCl		9.3		184
Poly fl⁵U	267	25	0.01 M Na citr., 0.1 M bor.	7.75	8.3		256
	S		0.15 M NaCl	7.6	8.1		157
Poly cl⁵U	S		0.15 M NaCl	8.3	8.35		157
Poly br⁵U	S		0.1 M NaCl		8.5	pH_m	211
	S		0.15 M NaCl	8.25	8.43		157
Poly io⁵U	S		0.15 M NaCl	8.5	8.64		157
Poly ho⁵U	S		0.15 M NaCl	8.5	8.3		157
Poly s⁴U	340		0.05 M NaCl	8.20	8.70		232
Poly s²,⁴U	S			7.5[d,r]	9.2		53b
Poly dU	S			9.30			58
Poly Q	S		0.15 M NaCl		9.05		157
					9.7		73
					9.7		200
Poly X	280		0.15 M NaCl	5.37	5.72;7.34		176

Polymer	Method	T °C	Medium	pK_{mono}	pK_{poly}	Note	Ref.
			1.3.2. _Copolymers_				
Poly d(A-T)		25	0.012 $\underline{M}$ Na$^+$, phosp.		11.2_8	co.	109
		25	0.5 $\underline{M}$ Na$^+$, phosp.		10.9_5	co.	109
		25	0.5 $\underline{M}$ Na$^+$, citr.		2.8-3.2	m	109
		260	0.15 $\underline{M}$ NaCl		3.25, 10.95		193, 194
		260	0.15 $\underline{M}$ NaCl		3.25 10.75	renatured (form II)	194
Poly d(A-s^{2}T)	245				11.04	co., rev.	141
Poly (A-U)					10.92	co.	209
Poly d(A-br^{5}U)		25	0.012 $\underline{M}$ Na$^+$, phosp.		9.73	co.	109
		25	0.5 $\underline{M}$ Na$^+$, phosp.		9.42	co.	109
		25	0.5 $\underline{M}$ Na$^+$, citr.		2.8-3.2	m	109
	S	25	0.1 $\underline{M}$ Na$^+$		10.1	pH_m	211
Poly (isoA-U)			0.15 $\underline{M}$ NaCl		10.75		178
Poly d(C,m^{7}G)	258		I = 0.02 $\underline{M}$		> 9.5	35%C^r	86
Poly (G,m^{1}G)			0.15 $\underline{M}$ NaCl		10.5		196
Poly (G ,sba^{8}G)	270		0.15 $\underline{M}$ NaCl		11.5	70%G;sco.	199
	400		0.15 $\underline{M}$ NaCl	7.1^n	7.7	nco.	199
Poly (U$_5$,s^{4}U)	330		0.05 $\underline{M}$ NaCl		8.9		221

Polymer	Method	T °C	Medium	pK_{mono}	pK_{poly}	Note	Ref.

1.3.3. Sugar-Phosphate Backbone Analogues

1.3.3.1. Polyvinylanalogues

Polymer	Method	T °C	Medium	pK_{mono}	pK_{poly}	Note	Ref.
Poly v[9]Acrid	420	22	I = 0.01 M;60% diox.	4.27	2.88	1	94
	420	22	I = 1.0 M;50% diox.		4.0	1	94
Poly v[1]Cyt	260, 275		25% p.g.	4.6	2.2		190
Poly v[1]Imid	225	25±0.2	0.01 M KCl		(5.4)		138
Poly v[4]Pyrid	P	36.6	0.04 M KCl;50% EtOH	5.2	(3.4)		139
Poly v[1]Ura	sc				10.6		133

Complex	Method	T °C	Medium	pK_{polym}	Note	Ref.

1.3.4. Complexes

Poly A Complexes

Complex	Method	T °C	Medium	pK_{polym}	Note	Ref.
Poly I	254	26	0.1 M NaCl	4.5;9.3	M_A=5.6;M_I=2.5	242
	248		0.15 M NaCl	9.38		176
	248		0.15 M NaCl, 0.01 M MgCl$_2$	9.14		176
	S		0.15 M NaCl	9.34		178
Poly U	S	20	0.01 M NaCl	5.12	A·U → A·A+U	41
	S	20	0.02 M NaCl	4.51	A·U → A·A+U	41
	S	20	0.03 M NaCl	4.35	A·U → A·A+U	41
	S	20	0.04 M NaCl	4.07	A·U → A·A+U	41
	S	20	0.08 M NaCl	3.71	A·U → A·A+U	41
	S	20	0.1 M NaCl	3.40	A·U → A·A+U	41
	S	20	0.1 M NaCl	4.20	A·U → A·2U+A·A	41
	S	20	0.3 M NaCl	4.40	A·U → A·2U+A·A	41
	S	20	0.6 M NaCl	4.40	A·U → A·2U+A·A	41
	S	20	0.8 M NaCl	4.40	A·U → A·2U+A·A	41
	S	20	1 M NaCl	4.40	A·U → A·2U+A·A	41
	S	20	I = 0.01 M	10.80	A·U → A+U	43
	S	20	I = 0.03 M	10.65	A·U → A+U	43
	S	20	I = 0.1 M	10.50	A·U → A+U	43
	S	20	I = 0.3 M	10.36	A·U → A+U	43
	S	20	I = 1 M	10.16	A·U → A+U	43
	S	20	I = 0.05 M	10.48		279
	P	20	I = 0.2 M	10.38		279
	S	20	I = 1.0 M	10.07		279
	S	20	I = 2.0 M	10.00		279

Complex	Method	T °C	Medium	pK_{polym}	Note	Ref.
Poly A Complexes (Con't)						
Poly U (Con't)						
	259	18.8	0.1 M KCl, 0.01 M acet. or carb.	(10.2)	M_A=22;M_U=7	241
		26.9	0.1 M KCl, 0.01 M acet. or carb.	(10.0)		241
		35.2	0.1 M KCl, 0.01 M acet. or carb.	(9.9)		241
	260	20	0.1 M NaCl	10.40		209
	260	VAR		VAR		45
2 Poly U	S		0.15 M NaCl	9.55;10.1		178
	S	20	0.2 M NaCl	3.09	A·2U → A·A+U	41
	S	20	0.3 M NaCl	2.82	A·2U → A·A+U	41
	S	20	0.4 M NaCl	2.70	A·2U → A·A+U	41
	S	20	0.6 M NaCl	2.60	A·2U → A·A+U	41
	S	20	0.8 M NaCl	2.49	A·2U → A·A+U	41
	S	20	1 M NaCl	2.35	A·2U → A·A+U	41
	260	22	acet. (< pH4), form. (> pH4)	(5.4)		2
Poly fl^{5}U	263	25	0.01 M Na citr., 0.1 M bor.	8.2	co.,rev.	256
Poly br^{5}U	260	24.5	0.1 M Na$^+$,10^{-3} mM EDTA	9.65	pH_m	211
2 Poly br^{5}U						
	260	26	0.10 M NaCl, 10^{-3} mM EDTA	9.38;9.68	pH_m	211
	285	26	0.10 M NaCl, 10^{-3} mM EDTA	9.38	pH_m	211
Poly (U$_5$,s^{5}U)						
	320		0.05 M NaCl	9.45		221
2 Poly X	250		0.15 M NaCl	8.76		176
	250		0.15 M NaCl,0.01 M MgCl$_2$	8.32		176

Complex	Method	T °C	Medium	pK_{polym}	Note	Ref.
Poly m^1A Complexes						
Poly X	250		0.15 M NaCl	8.54		176
	250		0.15 M NaCl, 0.01 M MgCl₂	9.50		176
Poly m^6A Complexes						
Poly X	250		0.15 M NaCl	8.24		176
Poly he^6A Complexes						
Poly X	250		0.15 M NaCl	8.42		176
Poly $A(CH_2O)$ Complexes						
Poly X	250		0.15 M NaCl	8.88		176
	250		0.15 M NaCl, 0.01 MgCl₂	8.42		176
Poly isoA Complexes						
Poly I	280		0.15 M NaCl	4.22;11.05	co.	178
Poly U	270		0.15 M NaCl	4.22;11.16	co.	178
2 Poly X	250		0.15 M NaCl	8.51		176
	250		0.15 M NaCl, 0.01 M MgCl₂	7.76		176
Poly dA Complexes						
2 Poly U	260	22	acet. (< pH4), form. (> pH4)	(4.3)		2
Poly d(A-C) Complexes						
Poly d(T-G)	256	22	0.02 M NaCl	11.9		283
Poly d(A-G) Complexes						
Poly d(T-C)	256	22	0.02 M NaCl	11.6		283

Complex	Method	T °C	Medium	pK_{polym}	Note	Ref.
Poly C and Poly C^+ Complexes						
Poly G						
	280		0.15 M NaCl	11.27	co.	180, 195, 196
	270,280	25		12.5	co.	81
Poly m^7G	S		0.15 M NaCl	>10	j	180
Poly (G,m^1G)			0.15 M NaCl	10.6		196
Poly $(G,m^{1,7}G)$			0.15 M NaCl	10.6		196
Poly (G,sba^8G) 270			0.15 M NaCl	12.0	co.	199
Poly I	P	25	0.05 M NaCl	3.7–4.0; 4.9		67
	S		0.15 M NaCl	10.05		180
	248		0.15 M NaCl	9.81		177
	P	VAR[Na$^+$] (0.05–1.0 M Na$^+$)		VAR		260
	P	VAR [Na$^+$]		VAR		261
	P	VAR [Na$^+$]		VAR	$c^+ \cdot I$	261
	P	VAR [Na$^+$]		VAR	$c \cdot c^+ \cdot I$	261
	P	VAR [Na$^+$] (0.05–1.0 M Na$^+$)		VAR	$c \cdot c^+ \cdot I$	260
Poly m^7I	S		0.15 M NaCl	8.02		180
Poly br^5C Complexes						
Poly G	280		0.15 M NaCl	12.56	co.	180, 195
Poly m^7G	S		0.14 M NaCl	>10	j	180
Poly I	248		0.15 M NaCl	10.43		175
	S		0.15 M NaCl	10.60		180

Complex	Method	T °C	Medium	pK$_{polym}$	Note	Ref.
Poly br^{5}C Complexes (Con't)						
Poly m^{7}I	S		0.15 M NaCl	9.21		180
Poly io^{5}C Complexes						
Poly I	248		0.15 M NaCl	10.49		175
Poly dC Complexes						
Poly dG	270	26	0.2 M NaCl, 1.5 mM K phosp.	11.4	rev.	205
	V	26	0.2 M NaCl, 1.5 mM K phosp.	11.1	irev.	205
Poly d(br^{5}C) Complexes						
Poly dG	270	26	0.2 M NaCl, 1.5 mM K phosp.	11.9	rev.	205
	V	26	0.2 M NaCl, 1.5 mM K phosp.	12	irev.	205
Poly I Complexes						
Poly X	245		0.15 M NaCl	6.12;7.34; 9.32		176
Poly U Complexes						
Poly X	265		0.15 M NaCl	6.50;10.18		176
	265		0.15 M NaCl,0.01 M MgCl$_2$	6.38;10.10		176

1.3.4.1. Complexes of Polyvinylanalogues

Complex	Method	Medium	pK$_{polym}$	Note	Ref.
Poly v^{1}Cyt Complexes					
Poly G	280	0.01 M NaCl, 5 mM Na bor.;25% p.g.	(12)	p	190

FOOTNOTES FOR SECTION 1

[a] pK values for uridine; see the reference for pK values of nucleic acid bases, nucleosides and nucleotides at $I = 0$ and 1 $\underline{M}$.

[b] Transition pH values were calculated according to various suppositions.

[c] pK determined from E_{max} and λ_{max} data.

[d] Nucleoside.

[e] Nucleoside diphosphate.

[f] Precipitates below pH 7.5.

[g] See ref. 57a.

[h] Phyrophosphate.

[i] Guanosine-2'(3')-monophosphate (product of alkaline hydrolysis of poly G).

[j] Alkaline degradation starts at pH 10.7.

[k] pK_a:uridine, 9.23; uridine-2'(3')-monophosphate, 9.39. All pK data from refs. 230 and 231 are accurate within +0.02.

[l] The pK_a of the monomer, 9-ethylacridine, was determined at 354 mμ; pK_a in water at 353 mμ was 6.10.

[m] Polymer is close to precipitation in the pH region indicated.

[n] pK for the modified monomer.

[o] Values obtained on titration with alkali; on backward titration with acid only a partial reversal of the hyperchromic shift was observed and a little lower values of transition pH were obtained.

^{p} Poly v^{1}C contained 10% 4-ethoxy-2-pyrimidinone due to hydrolysis.

^{q} pK$_a$:Guo (3.3;9.4), m^{7}Guo(6.7), m^{7}Guo-2'-P(7.0), m^{7}Guo-3'-P(6.9), m^{7}Guo-5'-P(7.1), m^{7}Guo-5'-PP(7.2), m^{7}Guo-5'-PPP(7.5), d(m^{7}Guo-5'-PPP)(7.5).

^{r} See ref. 53a; the pK's of uridine and 4-thiouridine are 9.3 and 8.2, respectively.

2. SPECTRAL DATA

INTRODUCTION

The molar absorptivity (molar extinction coefficient) E is defined as $E = A_\lambda \cdot mol^{-1} \cdot l \cdot cm^{-1}$, where A_λ is the absorbance at the wavelength λ and $mol \cdot l^{-1}$ refers to the molar concentration of polymeric or oligomeric phosphorus or base (this may be indicated in the Note column by letters P and B, respectively) and cm to the light path. Some E_λ values of oligomers are given as absorbance per 1 mol of oligomer; such values are underscored. Whenever possible, E_λ values are given for each listed compound together with hypochromicity or hyperchromicity data. Only in cases of uncommon derivatives are references included that give incomplete data (e.g., λ_{max}).

The λ columns give the wavelength of the absorption maximum at which E and hypochromicity (hyperchromicity) values were determined; wavelengths are marked by asterisks whenever λ is not the wavelength of the maximum or was not specified as such in the reference.

The h columns (oligonucleotides) give percent hypochromicity or hyperchromicity values (the latter are followed by the symbol "). When these values are given in the Note column (for polynucleotides), they are identified by abbreviations hpo. and hper., respectively; hyperchromicity accompanying the helix--coil transition is abbreviated as hper.(hx-c). Percent hypochromicity and hyperchromicity are alternatively defined as follows:

$$hpo. = \left[1 - \frac{E_{oligo}}{E_{mono}}\right] \cdot 100, \text{ or } \left[1 - \frac{E_{poly}}{E_{mono}}\right] \cdot 100,$$

$$\text{or } \left[1 - \frac{E_{complex}}{E_{poly}}\right] \cdot 100$$

$$\text{hper.} = \left[\frac{E_{mono}}{E_{oligo}} - 1\right]\cdot 100, \text{ or } \left[\frac{E_{mono}}{E_{poly}} - 1\right]\cdot 100,$$

$$\text{or } \left[\frac{E_{poly}}{E_{complex}} - 1\right]\cdot 100$$

$$\text{hper.(hx-c)} = \left[\frac{E_{coil}}{E_{helix}} - 1\right]\cdot 100$$

where E is the molar absorptivity (or absorbance) at the wavelength λ of the form specified by the subscript. The hypochromicity at a given wavelength is related to the hypochromism, H, of the absorption band, which is defined as $H(\%) = [1 - f_p/f_m)]\cdot 100$, where f_p and f_m are the oscillator strengths of the polymer and the equimolecular mixture of the component monomer, respectively. The reader should be cautioned that workers may report their results as thermal, residual, or total hyperchromicities. Thermal hyperchromicity is an increase in absorbance when the temperature is increased, usually from the temperature of existence of the ordered structure to that of the completely melted structure; such an increase in absorbance is often referred to as hyperchromicity of the helix--coil transition. Residual hyperchromicity is an increase in absorbance resulting from hydrolysis of the disordered (coiled, melted) polymer into its monomers. Total hyperchromicity is the sum of the thermal and residual hyperchromicities. The residual and total hypochromicities can be defined in analogous terms.

The temperature at which absorptivity, hypochromicity, and hyperchromicity values were obtained is given in the Note and $T(^{o}C)$ columns for oligonucleotides and polynucleotides, respectively. Absence of temperature data indicates that they were not given in the reference; temperature in such cases is usually understood as a room temperature.

Molar absorptivity values are generally determined by phosphate analysis or from the monomer concentration. The latter is determined upon hydrolysis of the polymer or oligomer from absorbance readings using literature E_λ values for the monomer. The method of determination of E_λ values may be indicated in the Note column by appropriate symbols listed under Abbreviations.

2.   SPECTRAL DATA

2.1.   OLIGONUCLEOTIDES

Oligomers	pH	Medium	λ mμ	E x 10^{-3} 1/mol·cm	h %	Note	Ref.

2.1.1.   Homooligomers

Oligomers	pH	Medium	λ mμ	E x 10^{-3} 1/mol·cm	h %	Note	Ref.
A-A	1	0.1 M HCl	257	28.4		230,0.83, 0.24, 0.05[a]	188
			257	28.4			150
		HClO$_4$	257	15.1	0.3	25°	280, 281
	4.9	0.1 M NaCl, 0.1 M acet.	257			20°	275
	6.93	0.08 M KClO$_4$, K phosp.; I = 0.1 M	258	13.9	9.4	25°	280, 281
	7.0	SSC	260*	12.7±0.07		d	223
	7.0	0.01 M phosp.	258	13.6	11.9	b	259
	7		260*		9.6	20°;a.h.	113
	N	0.01 M phosp.	258	13.6	11.9	rm. temp.; E$_{258}$(AMP)= 15.4	128
	(N)		258		12.5"		155
	7.4	0.1 M NaCl, 0.1 M Tris	257.0	13.8		4°	275
	7.4	0.1 M NaCl, 0.1 M Tris	258.5	14.7		65°	275
	7.4	0.1 M NaCl, 0.01 M Tris	257	13.6			20
	7.5	5 mM Tris	260*	11.36		25°	33
	7.5	7 M urea, 5 mM Tris	260*	11.61		25°	33
	8.5	0.08 M NaCl, 0.02 M buff.	257.5	13.6		22°	2
	8.5	0.08 M NaCl, 0.02 M buff.; 80% diox.	260	14.9		22°	2
	11.5	0.1 M KClO$_4$, KOH	258	13.9	9.4	25°	280, 281

Footnotes for Section 2.1 start on page 72.

Oligomers	pH	Medium	λ mμ	$E \times 10^{-3}$ 1/mol·cm	h %	Note	Ref.
A-A (Con't)							
		0.2 $\underline{N}$ NaOH	260*	13.3"		a.h.[c]	167, 169
		0.1 $\underline{N}$ NaOH	260*		10.7"	a.h.[c]	167
A2'-5'A	7.0	0.01 $\underline{M}$ phosp.	258.5	12.9	16.1	[b]	259
	N	0.01 $\underline{M}$ K phosp.	258	12.9	15.9	rm. temp.; E_{258}(AMP)= 15.4	128
	7.5	4.7 $\underline{M}$ KF,0.01 $\underline{M}$ Tris	258.5	12.9		a.h.	17
	8.5	0.08 $\underline{M}$ NaCl,0.02 $\underline{M}$ buff.	258	12.9		22°	2
		0.2 $\underline{N}$ NaOH	260*		18.4"	a.h.[c]	167, 169
			258		17"		155
(2'/3'-5)(pA-A)		0.01 $\underline{N}$ HCl	260*	14.25	0	a.h.[c]	166, 169
		0.01 $\underline{N}$ NaOH	258.5	13.1		E_{min}(229)= 2.9[c]	166
		0.2 $\underline{N}$ NaOH	260*		15.1"	a.h.[c]	166, 169
A5'-5'A		0.1 $\underline{M}$ HCl	257	13.6		Triethyl-ammonium salt	128
	N	0.025 $\underline{M}$ K phosp.	259	12.0	22.1	Triethyl-ammonium salt	128
A-Ap	7		260*		9.1	20°;a.h.	113
A-A2'p	7		260*		10.4	20°;a.h.	113
A5'pp5'A	1	0.1 M NaCl,0.02 $\underline{M}$ Tris	260*	12.5	13.6"		168
	7		260*	11.2	33.9"		168

Oligomers	pH	Medium	λ mμ	$E \times 10^{-3}$ 1/mol·cm	h %	Note	Ref.
A5'pp5'A (Con't)							
	7.0	SSC	260*	11.8±0.1		d	223
	(N)		260*		19.2"		155
	14		260*	11.2	33.9"		217
A5'ppp5'A	7.0	SSC	260*	12.0±0.07		d	223
A5'pppp5'A	6.0	SSC	260*	12.5±0.07		d	223
$_L A{-}_L A$	7.0	0.01 $\underline{M}$ phosp.	258	13.5	12.3	b	259
$_L A2'{-}5'_L A$	7.0	0.01 $\underline{M}$ phosp.	258.5	13.0	15.5	b	259
d(A-A)		H$_2$O	258	30.8		$\lambda_{max}(2)=256$	162
	N		260*	13.7	11.0	e	32
d(pA-A)	8.0	1 m$\underline{M}$ Tris	258	12.45		12.35,12.85[f]	34
			260*	11.8			25
	8.5	0.08 $\underline{M}$ NaCl,0.02 $\underline{M}$ buff.	257.5	12.9		22°	2
	8.5	0.08 $\underline{M}$ NaCl,0.02 $\underline{M}$ buff.;80% diox.	259.5	13.6		22°	2
Am-Ap	8.5	0.08 $\underline{M}$ NaCl,0.02 $\underline{M}$ buff.	258	12.9		22°	2
A-A-m	7		260*		9.3	20°;a.h.	113
A-A-A	6.80	0.1 $\underline{M}$ phosp.	260*	12.8		258,13.0[g]	30
	7		258	12.9			28
	7		260*	12.8(13.0)	16.9	H=7.9(8.8)%[h]	28
	7.4	0.1 $\underline{M}$ NaCl,0.01 $\underline{M}$ Tris	257.5	12.6			20
(2'/3'-5')(pA-A-A)							
		0.01 $\underline{N}$ HCl	260*	13.3	6.8"	a.h.[c]	166, 169

Oligomers	pH	Medium	λ $m\mu$	$E \times 10^{-3}$ $1/mol \cdot cm$	h %	Note	Ref.
(2'/3'-5')(pA-A-A) (Con't)							
		0.01 N NaOH	258	12.3		$E_{min}(229.5)$ = 3.0^c	166
		0.2 N NaOH	260*		22.8"	a.h.^{c}	166, 169
d(pA-A-A)	8.0	1 mM Tris	257	11.34		11.35;12.05^f	34
(Ap)$_6$A	4.5	0.1 M NaCl,0.05 M acet.	257				20
(Ap)$_7$A	4.5	0.1 M NaCl,0.05 M acet.	256				20
(A$_p$)$_{11}$A	4.5	0.1 M NaCl,0.05 M acet.	256				20
(Ap)$_{29}$A	4.5	0.1 M NaCl,0.05 M acet.	252				20
(Ap)$_{44}$A	4.5	0.1 M NaCl,0.05 M acet.	252				20
Poly A	4.5	0.1 M NaCl,0.05 M acet.	252	8.6			20
(Ap)$_n$A	7	0.1 M LiCl,0.01 M cacd.	259*	VAR		n = 1-6; VAR temp.	135
(Ap)$_3$A	(N)		260*	12.1			29
(Ap)$_4$A	(N)		260*	11.8			29
(Ap)$_5$A	(N)		260*	11.3			29
(Ap)$_4$A	7.4	0.1 M NaCl,0.01 M Tris	257.5	11.8			20
(Ap)$_5$A	7.4	0.1 M NaCl,0.01 M Tris	257	11.3			20
(Ap)$_6$A	7.4	0.1 M NaCl,0.01 M Tris	257	10.8			20
(Ap)$_7$A	7.4	0.1 M NaCl,0.01 M Tris	257	10.6			20

Oligomers	pH	Medium	λ mμ	$E \times 10^{-3}$ 1/mol·cm	h %	Note	Ref.
$(Ap)_{11}A$	7.4	0.1 $\underline{M}$ NaCl,0.01 $\underline{M}$ Tris	257	9.9			20
$(Ap)_{29}A$	7.4	0.1 $\underline{M}$ NaCl,0.01 $\underline{M}$ Tris	256.5	9.9			20
$(Ap)_{44}A$	7.4	0.1 $\underline{M}$ NaCl,0.01 $\underline{M}$ Tris	256.5				20
Poly A	7.4	0.1 $\underline{M}$ NaCl,0.01 $\underline{M}$ Tris	256.5	9.0			20
$(Ap)_nA$	7.4		max	VAR		n = 1-60; a.h.	20
$(A2'p)_nA$	7.4		max	VAR		n = 1-8; a.h.	20
$(pA)_7$	4.2	0.1 $\underline{M}$ NaClO$_4$;citr., phosp.; I = 0.12 $\underline{M}$	252	(9.7)		10.3^{o}	53
$(pA)_{10}$	4.2	0.1 $\underline{M}$ NaClO$_4$,citr., phosp.; I = 0.12 $\underline{M}$	257	(11.9)		55°	53
$(pA)_n$	Ac		max	VAR		n = 6-10; VAR temp.	53
$(2'/3'-5')(pA)_4$		0.01 $\underline{N}$ HCl	260*	12.54	13.2"	a.h.^{c}	166, 169
		0.01 $\underline{N}$ NaOH	258	11.5		$E_{min}(230)=$ 2.9^{c}	166
		0.2 $\underline{N}$ NaOH	260*		30.7"	a.h.^{c}	166, 169
$(2'/3'-5')(pA)_5$		0.01 $\underline{N}$ HCl	260*	12.22	16.4"	a.h.^{c}	166, 169
		0.01 $\underline{N}$ NaOH	258	11.4		$E_{min}(231)=$ 2.7^{c}	166
		0.2 $\underline{N}$ NaOH	260*		32.9"	a.h.^{c}	166, 169

Oligomers	pH	Medium	λ mμ	$E \times 10^{-3}$ 1/mol·cm	h %	Note	Ref.
d(pA)$_4$	8.0	1 mM Tris	257	11.02		10.87,12.23[f]	34
d(pA)$_5$	8.0	1 mM Tris	257	10.89		10.60,12.15[f]	34
d(pA)$_6$	8.0	1 mM Tris	257	10.32		10.40,11.99[f]	34
d(pA)$_7$	8.0	1 mM Tris	257	10.12		10.25,12.10[f]	34
d(pA)$_8$	8.0	1 mM Tris	257	9.98		10.15,11.83[f]	34
d(pA)$_9$	8.0	1 mM Tris	257	10.17		10.05,11.56[f]	34
d(pA)$_{10}$	8.0	1 mM Tris	257	9.91		10.00,11.61[f]	34
d(pA)$_{<300}$	8.0	1 mM Tris	257	9.39		9.45,12.04[f]	34
d(pA)$_6$			260*	10.2			25
d3'[(m^{6,6}An3')-(m^{6,6}Aacn3')]$_-$							
	2	0.1 M NaCl,0.02 M Tris	268	_32.9_	8.2"	235min[u]	168
	7		272	_30.2_	21.9"	237min[u]	168
	13		272	_30.2_	23.9"	237min[u]	168
C-C	1	0.1 M HCl	279	_26.0_		241,0.45, 2.02,1.50[e]	188, 150
	1	I = 0.1 M			0.1	25°	281
	4.0	0.1 M NaCl,0.05 M acet.	272.5	8.65		λ_{max}(CMP)=280	18
	7	0.1 M NaCl,0.05 M acet.	260*		7.8	20°;a.h.	113
	7	I = 0.1 M	max		7.2	25°	281
	7.5	4.7 M KF,0.01 M Tris	269	7.9		a.h.	17
	7.5	4.7 M KF,0.01 M Tris	269				19
	7.5	4.7 M KF,0.01 M Tris	269	7.92		E_{max}(272) CMP=8	18
	11.5	I = 0.1 M	max		7.2	25°	281
		0.2 N NaOH	260*		7.4"	a.h.[c]	167, 169

Oligomers	pH	Medium	λ mμ	$E \times 10^{-3}$ 1/mol·cm	h %	Note	Ref.
C2'-5'C	7.5	4.7 M KF,0.01 M Tris	269.5	7.8		a.h.	17
		0.2 N NaOH	260*		10.8"	a.h.[c]	167, 169
C-Cp	7	0.1 M NaCl,0.05 M acet.	260*	6.4		20°;a.h.	113
C2'-5'Cp		0.2 N NaOH	260*		11.0"	a.h.[c]	167
pC-C		0.2 N NaOH	260*		8.7"	a.h.[c]	167, 169
d(C-C)		H$_2$O	270	27.0		$\lambda_{max}(2)=279$	162
	N		260*	7.3	1.4	[e]	32
aC5'-5'dC	2		279	26.6		[i]	145
	5		271	17.0		[i]	145
a(C-C)		0.1 M NaCl,0.01 M Tris	271				19
C-C-C	7.5	4.7 M KF,0.01 M Tris	268	7.4			18
C2'-5'C2'-5'Cp		0.2 N NaOH	260*		15.7"	a.h.[c]	167
(Cp)$_5$C	4.0	0.1 M NaCl,0.05 M acet.	270.5	7.60			18
(Cp)$_6$C	4.0	0.1 M NaCl,0.05 M acet.	272.5	8.0			18
(Cp)$_7$C	4.0	0.1 M NaCl,0.05 M acet.	272.5	6.97			18
(Cp)$_8$C	4.0	0.1 M NaCl,0.05 M acet.	273	7.05			18
(Cp)$_9$C	4.0	0.1 M NaCl,0.05 M acet.	273	6.99			18
(Cp)$_{27}$C	4.0	0.1 M NaCl,0.05 M acet.	274	6.90			18

Oligomers	pH	Medium	λ mμ	E x 10^{-3} 1/mol·cm	h %	Note	Ref.
Poly C	4.0	0.1 M NaCl,0.05 M acet.	275	6.5			18
(Cp)$_3$C	7.5	4.7 M KF,0.01 M Tris	268	7.25			18
(Cp)$_4$C	7.5	4.7 M KF,0.01 M Tris	268.5	7.06			18
(Cp)$_5$C	7.5	4.7 M KF,0.01 M Tris	268.5	6.82			18
(Cp)$_6$C	7.5	4.7 M KF,0.01 M Tris	268.5	6.63			18
(Cp)$_7$C	7.5	4.7 M KF,0.01 M Tris	268.5	6.34			18
(Cp)$_8$C	7.5	4.7 M KF,0.01 M Tris	268.5	6.2			18
(Cp)$_9$C	7.5	4.7 M KF,0.01 M Tris	268.5	6.2			18
(Cp)$_{29}$C	7.5	4.7 M KF,0.01 M Tris	268	6.14			18
Poly C	7.5	4.7 M KF,0.01 M Tris	268	6.09			18
(C2'p)$_4$C	7.5	4.7 M KF,0.01 M Tris	269	7.60			18
(C2'p)$_5$C	7.5	4.7 M KF,0.01 M Tris	269	7.47			18
(pC)$_3$		0.2 N NaOH	260*		13.9"	a.h.[c]	167, 169
(pC)$_4$		0.2 N NaOH	260*		15.3"	a.h.[c]	167, 169
G-G	1	I = 0.1 M	max		0.5	25°	281
	7	I = 0.1 M	max		6.9	25°	281
	11.5	I = 0.1 M	max		-0.8	25°	281
G-Gp	7	0.1 M NaCl,0.05 M acet.	260*		8.4	20°;a.h.	113
d(G-G)	N		260*	10.8	6.1	[e]	32

Oligomers	pH	Medium	λ mμ	$E \times 10^{-3}$ 1/mol·cm	h %	Note	Ref.
d(G-G)(Con't)		H_2O	252	<u>27.4</u>		$\lambda_{max}(2)=255$	162
$(Gp)_n$	7.0	0.01 $\underline{M}$ phosp.		VAR		VAR n	83
I-Ip	7	0.1 $\underline{M}$ NaCl,0.05 $\underline{M}$ acet.	260*		6.0	$20°$;a.h.	113
T-T			267	<u>18.5</u>		p[k]	68
pT-T	8.0	0.1 $\underline{M}$ Tris	266	9.2		j	186
$(pT)_4$	8.0	0.1 $\underline{M}$ Tris	266	9.1		j	186
$(pT)_6$	8.0	0.1 $\underline{M}$ Tris	266	9.0		j	186
$(pT)_{10}$	8.0	0.1 $\underline{M}$ Tris	266	8.9		j	186
d(T-T)	2		267	<u>18.5</u>			162
	N		260*	8.4	3.7	e	32
	(N)		267	<u>18.5</u>		P	68
d(T-T-T)		0.01 $\underline{M}$ HCl	266	<u>25.2</u>		$E_{min}(235)=$ <u>6.6</u>	137
		H_2O	266	<u>25.4</u>		$E_{min}(235)=$ <u>8.2</u>	137
	N		266	<u>25.8</u>			68
		0.01 $\underline{M}$ NaOH	266	<u>22.0</u>		$E_{min}(247)=$ <u>15.0</u>	137
$d(pT)_2$	8.0	1 m$\underline{M}$ Tris	266	9.10		9.10[f]	34
$d(pT)_3$	8.0	1 m$\underline{M}$ Tris	266	9.05		8.90[f]	34
$d(pT)_4$	8.0	1 m$\underline{M}$ Tris	266	8.78		8.84[f]	34
$d(pT)_5$	8.0	1 m$\underline{M}$ Tris	266	8.75		8.78[f]	34

Oligomers	pH	Medium	λ mμ	$E \times 10^{-3}$ 1/mol·cm	h %	Note	Ref.
$d(pT)_6$	8.0	1 m$\underline{M}$ Tris	266	8.44		8.65^f	34
$d(pT)_7$	8.0	1 m$\underline{M}$ Tris	266	8.60		8.63^f	34
$d(pT)_8$	8.0	1 m$\underline{M}$ Tris	266	8.69		8.62^f	34
$d(pT)_9$	8.0	1 m$\underline{M}$ Tris	266	8.63		8.61^f	34
$d(pT)_{10}$	8.0	1 m$\underline{M}$ Tris	266	8.66		8.60^f	34
$d(pT)_{<300}$	8.0	1 m$\underline{M}$ Tris	265	8.54		8.49^f	34
U–U	1	0.1 $\underline{M}$ HCl	261	<u>19.3</u>		232,0.76, 0.35,0.05^a	188
	1	HClO$_4$	261	9.8	3.0	25°	280, 281
	6.93	0.08 $\underline{M}$ KClO$_4$, K phosp; I = 0.1 $\underline{M}$	261	9.8	1.7	25°	280, 281
	7.5	4.7 $\underline{M}$ KF,0.01 $\underline{M}$ Tris	260	9.1		a.h.	17
U–Up	7	0.1 $\underline{M}$ NaCl,0.02 $\underline{M}$ Tris	260*	9.62±0.05		20°;a.h.; E_{260}(UMP)= 10.00±0.05	231
	7	0.1 $\underline{M}$ NaCl,0.05 $\underline{M}$ acet.	260*		3.5	20°;a.h.	113
	11.5	0.1 $\underline{M}$ KClO$_4$,KOH	261	7.4	2.5	25°	280, 281
	11.6	0.1 $\underline{M}$ NaCl,0.02 $\underline{M}$ Tris	260*	7.35±0.06		20°,E_{260} (UMP)= 7.35±0.06	231
$(Up)_3$	7	0.1 $\underline{M}$ NaCl,0.02 $\underline{M}$ Tris	260*	9.46±0.05		20°;a.h.	231
	11.6	0.1 $\underline{M}$ NaCl,0.02 $\underline{M}$ Tris	260*	7.20±0.06		20°	231
$(Up)_4$	7	0.1 $\underline{M}$ NaCl,0.02 $\underline{M}$ Tris	260*	9.52±0.05		20°;a.h.	231
	11.6	0.1 $\underline{M}$ NaCl,0.02 $\underline{M}$ Tris	260*	7.33±0.06		20°	231
$(Up)_5$	7	0.1 $\underline{M}$ NaCl,0.02 $\underline{M}$ Tris	260*	9.44±0.05		20°;a.h.	231

Oligomers	pH	Medium	λ mμ	E x 10^{-3} 1/mol·cm	h %	Note	Ref.
(Up)$_5$(Con't)							
	11.6	0.1 M NaCl,0.02 M Tris	260*	7.29±0.06		20°	231
(Up)$_6$	7	0.1 M NaCl,0.02 M Tris	260*	9.36±0.05		20°;a.h.	231
	11.6	0.1 M NaCl,0.02 M Tris	260*	7.30±0.06		20°	231
(Up)$_7$	7	0.1 M NaCl,0.02 M Tris	260*	9.29±0.05		20°;a.h.	231
	11.6	0.1 M NaCl,0.02 M Tris	260*	7.15±0.06		20°	231
(Up)$_8$	7	0.1 M NaCl,0.02 M Tris	260*	9.28±0.05		20°;a.h.	231
	11.6	0.1 M NaCl,0.02 M Tris	260*	7.21±0.06		20°	231
(Up)$_9$	7	0.1 M NaCl,0.02 M Tris	260*	9.24±0.05		20°;a.h.	231
	11.6	0.1 M NaCl,0.02 M Tris	260*	7.19±0.06		20°	231
(Up)$_{10}$	7	0.1 M NaCl,0.02 M Tris	260*	9.27±0.05		20°;a.h.	231
	11.6	0.1 M NaCl,0.02 M Tris	260*	7.19±0.06		20°	231
(Up)$_{11}$	7	0.1 M NaCl,0.02 M Tris	260*	9.26±0.05		20°;a.h.	231
	11.6	0.1 M NaCl,0.02 M Tris	260*	7.19±0.06		20°	231
Poly U	7	0.1 M NaCl,0.02 M Tris	260*	9.10±0.05		20°;a.h.	231
	11.6	0.1 M NaCl,0.02 M Tris	260*	7.06±0.06		20°	231

Oligomers	pH	Medium	λ mμ	$E \times 10^{-3}$ 1/mol·cm	h %	Note	Ref.

2.1.2. Heterooligomers

Oligomers	pH	Medium	λ mμ	$E \times 10^{-3}$ 1/mol·cm	h %	Note	Ref.
A–C	1	HClO$_4$	266	10.6	1.8	25°	280, 281
	1	0.1 $\underline{M}$ HCl	266	20.2		234,0.72, 0.76,0.50[a]	188
	6.93	0.08 $\underline{M}$ KClO$_4$, K phosp.; I = 0.1 $\underline{M}$	261	10.5	7.6	25°	280, 281
	N	0.01 $\underline{M}$ K phosp.	260	10.6	6.8	rm. temp.[b]	128
	7.5	4.7 $\underline{M}$ KF,0.01 $\underline{M}$ Tris	261	10.0		a.h.	17
	7.5	4.7 $\underline{M}$ KF,0.01 $\underline{M}$ Tris	261				19
	8.6	0.02 $\underline{M}$ (NH$_4$)HCO$_3$,0.01 $\underline{M}$ MgCl$_2$	260*	21.3	5"	s.v.h.	236
	11.5	0.1 $\underline{M}$ KClO$_4$,KOH	261	10.4	9.0	25°	280, 281
		0.2 $\underline{N}$ NaOH	260*		7.0"	a.h.[c]	166, 167
A2'–5'C	N	0.01 $\underline{M}$ K phosp.	260	10.1	11	rm. temp.[b]	128
	7.5	4.7 $\underline{M}$ KF,0.01 $\underline{M}$ Tris	262	10.0		a.h.	17
A–Cp		0.1 $\underline{N}$ HCl	260*	21.2		267,236, 0.73,0.80, 0.50[1]	236
		0.1 $\underline{N}$ NaOH	260		8.0"	a.h.[c]	167
A–aC	7.5	4.7 $\underline{M}$ KF,0.01 $\underline{M}$ Tris	261				19
d(A–C)	N		260*	10.6	7.1	[a]	32
		H$_2$O	262	22.2		λ_{max}(2)=265	162
A–G	1	0.1 $\underline{M}$ HCl	257	25.8		229,0.86, 0.43,0.25[a]	188
			257	25.4			150
	1	I = 0.1 $\underline{M}$	max		3.0	25°	281
	7	I = 0.1 $\underline{M}$	max		5.8	25°	281

Oligomers	pH	Medium	λ mμ	$E \times 10^{-3}$ 1/mol·cm	h %	Note	Ref.
A-G (Con't)							
	7.5	5 mM Tris	260*	10.58		25°	33
	7.5	7 M urea, 5 mM Tris	260*	10.48		25°	33
	8.6		260*	25		255(257), 0.98(0.92), 0.38(0.41), 0.13(0.22)[n]	284
	11.5	I = 0.1 M	max		4.0	25°	281
A-Gp	1	0.1 M HCl	258	25.3		m	4
	1	0.1 M HCl	260*	25.0		5.8(229), 0.85,0.44, 0.26	4
	7	0.05 M phosp.	257	25.0		m	4
	7	0.05 M phosp.	260*	24.4	9.4	5.7(226), 0.93,0.40, 0.17[m]	4
	13	0.1 M KOH	261	24.9		m	4
	13	0.1 M KOH	260*	25.0		7.1(231), 0.79,0.41, 0.08[m]	4
d(A-G)	N		260*	12.5	10.0	e	32
		H_2O	253	27.3		$\lambda_{max}(2)=256$	162
d(A-T)	N		260*	11.4	4.8	e	32
		H_2O	261	20.5		$\lambda_{max}(2)=261$	162
A-U	1	0.1 M HCl	258	23.4		230,0.80, 0.24,0.04[a]	188
			258	23.2			150
	1	$HClO_4$	258	12.1	3.0	25°	280, 281
	6.93	0.08 M $KClO_4$, K phosp.; I = 0.1 M	260	12.0	5.0	25°	280, 281

Oligomers	pH	Medium	λ mμ	$E \times 10^{-3}$ 1/mol·cm	h %	Note	Ref.
A-U (Con't)							
	7.5	4.7 $\underline{M}$ KF,0.01 $\underline{M}$ Tris	260	11.4		a.h.	17
	8.6	0.02 $\underline{M}$ (NH$_4$)HCO$_3$, 0.01 $\underline{M}$ MgCl$_2$	260*	23.4	5"	s.v.h.	236
	11.5	0.1 $\underline{M}$ KClO$_4$,KOH	260	11.1	3.8	25°	280, 281
		0.2 $\underline{N}$ NaOH	260*		4.1"	a.h.	167, 169
A2'-5'U		0.01 $\underline{N}$ HCl	260*		12.3"	a.h.^{c}	167, 169
		0.01 $\underline{N}$ HCl	258	21.8		E_{min}(231)= 5.6^c	166
	7		260*		10.0"	a.h.	169
		H$_2$O	260*		16.0"	a.h.^{c}	166
		0.01 $\underline{N}$ NaOH	261	20.0		E_{min}(236)= 8.8^c	166
		0.2 $\underline{N}$ NaOH	260*		12.8"	a.h.^{c}	166
		0.01 $\underline{N}$ HCl	260*		13.6"	a.h.	169
		0.2 $\underline{N}$ NaOH	260*		10.3"	a.h.	169
A-Up		0.1 $\underline{N}$ NaOH	260*		4.1"	a.h.^{c}	167
A 5'pp5'U	1	0.1 $\underline{M}$ NaCl,0.02 $\underline{M}$ Tris	260*	11.1	9.0"		168
	7		260*	11.1	12.6"		168
	14		260*	10.5	6.7"		168
A-s^{4}U	5.8		259; 333	(17.4); 17.9			115
	5.8		259	16.9	11.4 (260)	289(3.4)^P	114
	5.8		333	17.9	14.0 (330)		114
	10.0		261	17.7		287(7.0)^P	114
	10.0		318	17.7			114

Oligomers	pH	Medium	λ mμ	E x 10^{-3} 1/mol·cm	h %	Note	Ref.
A-Q	8.6		260*	22.5		260(261), 0.77(0.79), 0.30(0.42), 0.07(0.30)°	55
C-A	1	0.1 M HCl	266	20.8		234,0.72, 0.77,0.50[a]	188
			266	20.6			150
	1	I = 0.1 M	max		-1.0	25°	281
	7	I = 0.1 M	max		7.8	25°	281
	7.5	4.7 M KF,0.01 M Tris	261.5	10.0		a.h.	17
	7.5	4.7 M KF,0.01 M Tris	261.5				19
	11.5	I = 0.1 M	max		6.5	25°	281
		0.2 N NaOH	260*		10.2"	a.h.[c]	167, 169
C2'-5'A	7.5	4.7 M KF,0.01 M Tris	269.5	7.78			18
		0.2 N NaOH	260*		9.7"	a.h.[c]	167, 169
aC-A	7.5	4.7 M KF,0.01 M Tris	261.5				19
d(C-A)	N		260*	10.6	7.1	[e]	32
		H$_2$O	262	22.2		λ_{max}(2)= 267	162
C-G	1	I = 0.1 M	max		9.8	25°	281
	7	I = 0.1 M	max		3.2	25°	281
	8.6		260*	18.2		253(278), 1.04(0.78), 0.76(1.14), 0.30(0.77)[n]	55
	11.5	I = 0.1 M			9.2	25°	33
			276	19.8			150
C-Gp	1	0.1 M HCl	278	19.0		[m]	4

Oligomers	pH	Medium	λ mμ	E x 10^{-3} 1/mol·cm	h %	Note	Ref.
C-Gp (Con't)							
	1	0.1 M HCl	260*	17.1		6.0(234), 0.73,1.11, 0.82 m	4
	7	0.05 M phosp.	256	18.1		m	4
	7	0.05 M phosp.	260*	17.6	6.6	9.7(225), 1.00,0.72, 0.32m	4
	13	0.1 M KOH	268	18.2		m	4
	13	0.1 M KOH	260*	17.3		11.1(233), 0.85,0.76, 0.25m	4
$Cm^{2,2}G{>}p$	8.6		260*	20		262(268), 0.80(0.55), 0.67(1.05), 0.45(0.77)n	284
d(C-G)	N		260*	9.0	5.3	e	32
		H_2O	254	19.7		$\lambda_{max}(2){=}277$	162
d(pC-G)			255	19.9			59
aC-N	2;5		max			i	282
aC-dN	2;5		max			i	282
dC-T	2		273	21.2		k	68
	2		273	21.2			162
	N		260*	7.6	5.8	e	32
		H_2O	268	18.8			162
C-U	1	0.1 M HCl	269	18.0		234,0.63 0.99,0.61[a]	188
			269	18.0			150
	1	I = 0.1 M	max		2.7	25°	281
	7	I = 0.1 M	max		6.3	25°	281

Oligomers	pH	Medium	λ mμ	$E \times 10^{-3}$ 1/mol·cm	h %	Note	Ref.
C-U (Con't)							
	7.5	4.7 $\underline{M}$ KF,0.01 $\underline{M}$ Tris	264.5	8.1		a.h.	17
	7.5	4.7 $\underline{M}$ KF,0.01 $\underline{M}$ Tris	264.5				19
	8.6		260*	16.5		265(269), 0.76(0.64), 0.65(0.94), 0.17(0.52)^n	55
	11.5	I = 0.1 $\underline{M}$	max		4.0	25°	33
		0.01 $\underline{N}$ NaOH	260*		8.7"	a.h.[c]	167
C2'-5'U		0.01 $\underline{N}$ NaOH	260*		7.9"	a.h.[c]	167
C-Up	8.6		260*	16.5		265(269), 0.76(0.64), 0.65(0.94), 0.17(0.52)^n	55
aC-U	7.5	4.7 $\underline{M}$ KF,0.01 $\underline{M}$ Tris	266				19
m^{5}C-G	8.6		260*	16		250(281), 0.96(0.86), 1.00(1.20),0.71 (1.00)^n	284
G-A	1	0.1 $\underline{M}$ HCl	257	25.0		230,0.86, 0.42,0.25[a]	188
			257	25.4			150
	1	I = 0.1 $\underline{M}$	max		3.7	25°	281
	7	I = 0.1 $\underline{M}$	max		7.6	25°	281
	7.5	4.7 $\underline{M}$ KF,0.01 $\underline{M}$ Tris	258	13.1		a.h.	17
	11.5	I = 0.1 $\underline{M}$	max		5.8	25°	33
		0.2 $\underline{N}$ NaOH	260*		3.3"	a.h.[c]	167, 169
G2'-5'A		0.2 $\underline{N}$ NaOH	260*		5.0"	a.h.[c]	167, 169
d(G-A)	N		260*	12.6	9.3"	[c]	32
		H$_2$O	255	27.3		$\lambda_{max}(2)=257$	162

Oligomers	pH	Medium	λ mμ	E x 10^{-3} 1/mol·cm	h %	Note	Ref.
G-C	1	0.1 M HCl	276	19.8		235,0.76, 1.11,0.79[a]	188
			276	19.8			150
	1	HClO$_4$	277	9.4	10.7	25°	280, 281
	6.93	0.08 M KClO$_4$,K phosp.; I = 0.1 M	255	9.1	8.7	25°	280, 281
	7.5	4.7 M KF,0.01 M Tris	266	8.9		a.h.	17
	8.6	0.02 M (NH$_4$)HCO$_3$,0.01 M MgCl$_2$	260*	18.4	4"	s.v.h.	236
	11.5	0.1 M KClO$_4$,KOH	267	9.3	6.0	25°	280, 281
		0.2 N NaOH	260*		3.0"	a.h.[c]	167, 169
G-Cp		0.1 N HCl	260*	19.2		278,235, 0.76,1.12, 0.74[l]	276
	8.6		260*	18		255(276), 0.98(0.82), 0.72(1.08), 0.31(0.71)[n]	284
		0.2 N NaOH	260*		3.8"	a.h.[c]	167, 169
G-m^{5}Cp	8.6		260*	17		253(281), 1.07(0.86), 0.91(1.17), 0.56(0.95)[n]	284
d(G-C)	N		260*	8.8	7.4	e	32
		H$_2$O	256	19.7			162
d(G-T)		H$_2$O	256	20.8		λ_{max}(2)=256	162
	N		260*	10.0	1.8	e	32
G-U	1	0.1 M HCl	257	20.5		229,0.84, 0.52,0.30[a]	188

Oligomers	pH	Medium	λ $m\mu$	$E \times 10^{-3}$ 1/mol·cm	h %	Note	Ref.
G-U (Con't)							
	1	HClO$_4$	259	10.3	6.2	25°	280, 281
	6.93	0.08 M KClO$_4$,K phosp.; I = 0.1 M	256	10.9	2.4	25°	280, 281
	7.5	4.7 M KF,0.01 M Tris	258	10.0		a.h.	17
	8.6	0.02 M (NH$_4$)HCO$_3$,0.01 M MgCl$_2$	260*	20.3	5"	s.v.h.	236
	11.5	0.1 M KClO$_4$,KOH	261	9.1	4.3	25°	280, 281
G-Up		0.1 N HCl	260*	27.6		259,230, 0.82,0.53, 0.30[l]	276
	8.6		260*	20		256(258), 0.96(0.90), 0.52(0.54), 0.19(0.30)[n]	284
G-s^{4}U	6.0		252, 333	(18.6), 17.0			115
	6.0		252	17.3	3.6 (260)	297(8.4)[p]	114
	6.0		333	17.9	14.3 (330)		114
	10.0		255	16.1		294 (9.9)[p]	114
	10.0		318	18.0			114
m^{2,2}G-Qp		0.1 N HCl	260*	23.8		263,236, 0.70,0.57, 0.41[l]	276
s^{6}G-U	6.0		280, 342				115
	10.0		252, 320				115
d(T-A)	N		260*	11.7	2.4	e	32
		H$_2$O	260*	20.5		λ_{max}(2)= 260	162

Oligomers	pH	Medium	λ mμ	$E \times 10^{-3}$ 1/mol·cm	h %	Note	Ref.
d(T-A) (Con't)			261	20.2			68
d(T-C)	2		273	21.2			68
	N		260*	8.1	0.0	e	32
		H_2O	268	18.8		$\lambda_{max}(2)=273$	162
d(T-G)	N		260*	9.5	7.4	e	32
		H_2O	255	20.8		$\lambda_{max}(2)=257$	162
d(pT-G)			257	20.8			59
T-Q	8.6		260*	15.9		263(269), 0.70(0.75), 0.57(0.84), 0.15(0.45)°	55
T-Qp	8.6		260*	15.9		263(269), 0.70(0.75), 0.57(0.84), 0.15(0.54)°	55
U-A	1	0.1 M HCl	257	23.0		230,0.82, 0.29,0.05[a]	188
	1	I = 0.1 M	max		3.3	25°	281
	7	I = 0.1 M	max		3.0	25°	281
	11.5	I = 0.1 M	max		1.5	25°	281
		0.2 N NaOH	260*		3.2"	a.h.[c]	167, 169
			258	23.2			150
U2'-5'A		0.2 N NaOH	260*		5.2"	a.h.[c]	167, 169
U-C	1	0.1 M HCl	269	20.8		235,0.63, 0.98,0.60[a]	188
			269	18.0			150

Oligomers	pH	Medium	λ mμ	E x 10^{-3} 1/mol·cm	h %	Note	Ref.
U-C (Con't)							
	1;7; 11.5	I = 0.1 M	max		3.2; 2.4; 1.8	25°	281
U-aC	7.5	4.7 M KF,0.01 M Tris	263				17
U-G	1	0.1 M HCl	258	20.5		229,0.84, 0.52,0.30[a]	188
			258	20.5			150
	1;7; 11.5		258	20.5	5.6; 5.3; 4.2	25°	281
	7.5	4.7 M KF,0.01 M Tris	255	10.5		a.h.	17
	8.6		260*	20.6		258(256), 0.90(0.96), 0.47(0.49), 0.16(0.12)[n]	55
dU-G	2		258			230,0.93, 0.48,0.19[a]	93
U-Gp	1	0.1 M HCl	260	20.6		4.80(280), 0.80,0.54, 0.30[m]	4
	7	0.05 M phosp.	257	21.0		m	4
	7	0.05 M phosp.	260*	20.4	5.0	5.5(227), 0.94,0.53, 0.21[m]	4
	13	0.1 M KOH	262	18.0		m	4
	13	0.1 M KOH	260*	18.0		10.5(235), 0.84,0.52, 0.11[m]	4
Um-G	8.6		260*	20.6		256(258), 0.97(0.92), 0.54(0.55), 0.24(0.28)[n]	55
A-A-C	1		259	11.3		H=0.1 (-0.7)%[h]	28
	1		260*	11.3 (11.7)	4.5	h	28

2. SPECTRAL DATA

Oligomers	pH	Medium	λ mμ	E x 10^{-3} 1/mol·cm	h %	Note	Ref.
A-A-C (Con't)							
	1		260*		2.0		237
	1.08	HClO$_4$	260*	11.3		259,11.7[g]	30
	6.80	0.1 M phosp.	260*	10.2		259,10.9[g]	30
	7		259	10.2		H=11.1 (9.5)%[h]	28
	7		260*	10.2 (10.9)	19.9	[h]	28
	7		260*		18.6		237
	8.6	0.02 M (NH$_4$)HCO$_3$, 0.01 M MgCl$_2$	260*	<u>31.6</u>	18"		236
	11.42	0.1 M KClO$_4$,NaOH	260*	10.4		255,10.9[g]	30
	11.5		260*		18.4		237
	11.5		259	10.4		H=8.2 (12.6)%[h]	28
	11.5		260*	10.4 (10.9)	17.6	[h]	28
		0.2 N NaOH	260*		23.3"	a.h.[c]	167, 169
A-A-Cp	8.6		260*	<u>31</u>		260(260), 0.80(0.78), 0.43(0.55), 0.17(0.31)[n]	284
		0.2 N NaOH	260*		22.5"	a.h.[c]	167, 169
A-A-Gp	1	0.1 M HCl	258	<u>39.7</u>		[m]	4
	1	0.1 M HCl	260*	<u>39.4</u>		<u>10.8</u>(231), 0.83,0.37, 0.18[m]	4
	1	0.1 M NaCl,HClO$_4$	258	<u>39.7</u>	1.9 (260)		112
	7	0.05 M phosp.	258	<u>35.3</u>		[m]	4
	7	0.05 M phosp.	260*	<u>34.8</u>	17.9	<u>9.2</u>(228), 0.59,0.36, 0.12[m]	4
	7	0.1 M NaCl	258	<u>35.3</u>	17.9 (260)		112

Oligomers	pH	Medium	λ mμ	E x 10^{-3} 1/mol·cm	h %	Note	Ref.
A–A–U	1		257.5	12.6		H=1.0 (1.2)%[h]	28
	1		260*	12.4 (12.8)	5.6	[h]	28
	1		260*		5.8		237
	1.08	HClO$_4$	260*	12.4		257.5, 12.8[g]	30
	6.80	0.1 M̲ phosp.	260*	11.3		258.5, 11.9[g]	30
	7		258.5	11.4		H=7.3 (5.9)%[h]	28
	7		260*	11.3 (11.9)	16.4	[h]	28
	7		260*		15.5		237
	8.6	0.02 M̲ (NH$_4$)HCO$_3$, 0.01 M̲ MgCl$_2$	260*	34.8	14"	s.v.h.	236
	11.42	0.1 M̲ KClO$_4$	260*	11.0		258,114[g]	30
	11.5		260*		11.9		237
	11.5		258	11.1		H=6.8 (7.5)%[h]	28
	11.5		260*	11.0 (11.4)	14.8	[h]	28
A–C–Gp	1	0.1 M̲ HCl	261	30.2		[m]	4
	1	0.1 M̲ HCl	260*	30.1		10.1(233), 0.76, 0.69, 0.44[m]	4
	1	0.1 M̲ NaCl,KClO$_4$	261	30.2	8.2 (260)		112
	7	0.05 M̲ phosp.	259	29.6		[m]	4
	7	0.05 M̲ phosp.	260*	29.6	13.6	12.3(228), 0.86, 0.49, 0.20[m]	4
	7	0.1 M̲ NaCl	259	29.6	13.6 (260)		112
	13	0.1 M̲ KOH	262	29.7		[m]	4

Oligomers	pH	Medium	λ mμ	E x 10^{-3} 1/mol·cm	h %	Note	Ref.
A-C-Gp (Con't)							
	13	0.1 M KOH	260*	29.6		14.6(232), 0.81,0.49, 0.14[m]	4
	13	0.1 M NaCl,NaOH	262	29.7	13.8 (260)		112
			260*		15.7		111
A-G-C	1		260*		1.1		237
	7		260*		9.6		237
	8.6	0.02 M (NH$_4$)HCO$_3$, 0.01 M MgCl$_2$	260*	30.6	11"	s.v.h.	236
	11.5		260*		6.1		237
A-G-U	1		258	11.7		H=2.9 (4.3)%[h]	28
	1		260*	11.6(11.5)	4.4	h	28
	1		260*		3.3		237
	1.08	HClO$_4$	260*	11.6		258,11.5[g]	30
	6.80	0.1 M phosp.	260*	11.3		257,11.6[g]	30
	7		257	11.5		H=3.9 (2.6)%[h]	28
	7		260*	11.3(11.6)	7.9	h	28
	7		260*		6.7		237
	8.6	0.02 M (NH$_4$)HCO$_3$, 0.01 M MgCl$_2$	260*	33.3	9"	s.v.h.	236
	11.42	0.1 M KClO$_4$,NaOH	260*	11.3		259,10.9[g]	30
	11.5		260*		2.6		237
	11.5		259	11.3		H=-2.5 (-1.0)%[h]	28
	11.5		260	11.3 (10.9)	2.3	h	28
A-U-G			260*		10.3		111
A-U-Gp	1	0.1 M HCl	260	34.0		8.6(231), 0.80,0.41, 0.20[m]	4

Oligomers	pH	Medium	λ mμ	$E \times 10^{-3}$ 1/mol·cm	h %	Note	Ref.
A-U-Gp (Con't)							
	1	0.1 $\underline{M}$ NaCl,HClO$_4$	260	<u>34.0</u>	4.8(260)		112
	7	0.03 $\underline{M}$ phosp.	259	<u>33.6</u>		m	4
	7	0.05 $\underline{M}$ phosp.	260*	<u>33.5</u>	9.3	8.4(228)0.85, 0.39, 0.13^m	4
	7	0.1 $\underline{M}$ NaCl	259	<u>33.6</u>	9.3(260)		112
	13	0.1 $\underline{M}$ KOH	261	<u>31.9</u>		m	4
	13	0.1 $\underline{M}$ KOH	260*	<u>31.8</u>		14.3(233),0.80, 0.39,0.07^m	4
	13	0.1 $\underline{M}$ NaCl,NaOH	261	<u>31.9</u>	6.5(260)		112
C-A-G	8.6		260*	<u>30</u>		257(260), 0.94(0.82), 0.53(0.72), 0.20(0.45)^n	284
C-A-Gp	1	0.1 $\underline{M}$ HCl	259	<u>30.4</u>		m	4
	1	0.1 $\underline{M}$ HCl	260*	<u>30.4</u>		8.5(231),0.80, 0.55,0.34^m	4
	1	0.1 $\underline{M}$ NaCl,HClO$_4$	259	<u>30.4</u>	7.4(260)		112
	7	0.05 $\underline{M}$ phosp.	258	<u>30.0</u>			4
	7	0.05 $\underline{M}$ phosp.	260*	<u>29.7</u>	13.3	9.6(227),0.90, 0.44,0.18^m	4
	7	0.1 $\underline{M}$ NaCl	258	<u>30.3</u>	13.1(260)		112
	13	0.1 $\underline{M}$ KOH	261	<u>30.4</u>		m	4
	13	0.1 $\underline{M}$ KOH	260*	<u>30.4</u>		11.9(231),0.80, 0.44,0.11^m	4
	13	0.1 $\underline{M}$ NaCl, NaOH	261	<u>30.4</u>	11.5(260)		112
			260*		15.3		111
C-A-U		0.2 $\underline{N}$ NaOH	260*		12.5"	a.h.^{c}	167,169
C-C-A	8.6		260*	<u>27</u>		265(270),0.80 (0.64),0.55(1.07), 0.18(0.69)^n	284
C-C-G	8.6		260*	<u>24</u>		255(277),0.98 (0.71),0.78(1.26), 0.32(0.89)^n	284

Oligomers	pH	Medium	λ mμ	$E \times 10^{-3}$ 1/mol·cm	h %	Note	Ref.
C-C-Gp	1	0.1 M HCl	278	30.6		m	4
	1	0.1 M HCl	260*	22.8		8.2 (237), 0.64, 1.33, 0.98[m]	4
	1	0.1 M NaCl, HClO$_4$	278	30.6	9.9(260)		112
	7	0.05 M phosp.	267	23.3		m	4
	7	0.05 M phosp.	260*	23.3	11.2	16.7 (228), 0.93, 0.77, 0.36[m]	4
	7	0.1 M NaCl	267	23.3	11.2(260)		112
	13	0.1 M KOH	269	25.0		m	4
	13	0.1 M KOH	260*	23.1		18.2 (236) 0.87, 0.80 0.30[m]	4
	13	0.1 M NaCl, NaOH	269	25.0	14.5(260)		112
C-ac^{6}C-G	8.6		260*	24		250(255, 280),1.20 (1.00), 0.71 (1.02) 0.44 (0.86)[n]	284
C-U-A		0.2 N NaOH	260*		7.5"	a.h.[c]	167, 169
G-A-br^{8}A		H$_2$O	259.5				104
G-A-C	1		260*		3.0		237
	7		260*		10.8		237
	8.6	0.02 M (NH$_4$)HCO$_3$, 0.01 M MgCl$_2$	260*	30.4	12"	s.v.h.	236
	11.5		260*		8.4		237
G-A-U	1		258.5	11.3		H=7.7 (3.3)%[h]	28
	1		260*	11.2(11.7)	7.9	h	28
	1		260*		5.2		237
	1.08	HClO$_4$	260*	11.2		258.5[g]	30
	6.80	0.1 M phosp.	260*	11.0		257.5[g]	30

Oligomers	pH	Medium	λ mμ	E x 10^{-3} 1/mol·cm	h %	Note	Ref.
G-A-U (Con't)							
	7		257.5	11.1		H=8.9 (6.2)%[h]	28
	7		260*	11.0(11.3)	10.6	[h]	28
	7		260*		9.5		237
	8.6	0.02 $\underline{M}$ (NH$_4$) HCO$_3$, 0.01 $\underline{M}$ MgCl$_2$	260*	<u>33.5</u>	8"	s.v.h.	236
	11.42	0.1 $\underline{M}$ KClO$_4$, NaOH	260*	10.8		259.5[g]	30
	11.5		260*		4.9		237
	11.5		259.5	10.8		H=2.0 (2.1)%[h]	28
	11.5		260*	10.8(10.8)	6.6	[h]	28
		0.02 $\underline{N}$ NaOH	260*		11.3"	a.h.[c]	167, 169
G-G-C	1		260	9.9	1.0	H=4.2%	28
	1		260		4.7		237
	1.08	HClO$_4$	260*	9.9			30
	6.80	0.1 $\underline{M}$ phosp.	260*	9.2		255[g]	30
	7		255	9.7		H=13.0%	28
	7		260*	9.2	9.8		28
	7		260*		7.9		237
	8.6	0.2 $\underline{M}$ (NH$_4$) HCO$_3$ 0.01 $\underline{M}$ MgCl$_2$	260*	<u>30.2</u>	7"	s.v.h.	236
	11.42	0.1 $\underline{M}$ KClO$_4$, NaOH	260*	9.4		266[g]	30
	11.5		260*		2.3		237
	11.5		266	9.6		H=6.0%	28
	11.5		260*	9.4	5.7		28
G- G-Cp	8.6		260*	<u>29</u>		254(259), 1.04 (0.90), 0.68 (0.87), 0.34 (0.59)[n]	284
pG-G-Cp	8.6		260*	<u>29</u>		255(258),1.03 (0.95),0.73 (0.84), 0.38 (0.58)[n]	284
G-G-U	1		258	11.1		H=1.4%	28

Oligomers	pH	Medium	λ mμ	$E \times 10^{-3}$ 1/mol·cm	h %	Note	Ref.
G-G-U (Con't)							
	1		260*	11.0	1.8		28
	1		260*		5.4		237
	1.08	HClO$_4$	260*	11.0		258^g	30
	6.80	0.1 $\underline{M}$ phosp.	260*	10.5		255^g	30
	7		255	11.2		H=5.2%	28
	7		260*	10.5	4.4		28
	7		260*		6.0		237
	8.6	0.02 $\underline{M}$ (NH$_4$) HCO$_3$ 0.01 $\underline{M}$ MgCl$_2$	260*	$\underline{31.6}$	4"	s.v.h.	236
	11.42	0.1 $\underline{M}$ KClO$_4$ NaOH	260*	10.6		261.5^g	30
	11.5		260*		-5.6		237
	11.5		261.5	10.6		H=5.6%	28
	11.5		260	10.6	-3.9		28
G-U-A		0.2 $\underline{N}$ NaOH	260*		5.3"	a.h.^{c}	167, 169
G$_m$-G-hUp	8.6		260*	$\underline{23}$		253(256), 1.13(0.99), 0.63(0.62), 0.30(0.40)^n	284
d(pT-C-A)	2		265	$\underline{33.0}$		0.75, 0.75^r	187
d(T-C-C)	2		275	$\underline{31.5}$		0.55,1.49^r	187
d(pT-C-C)	2		276	$\underline{32.0}$		0.52, 1.54^r	187
d(pT-G-G)			256	$\underline{34.2}$			59
d(pT-G-T)			261	$\underline{29.4}$			59
T-Q-C	8.6		260*	$\underline{21.2}$		270(266),0.61 (0.74),0.94 (0.69),0.49 (0.23)^o	55
d(pT-T-A)			261	$\underline{33.0}$			59

Oligomers	pH	Medium	λ mμ	E x 10^{-3} 1/mol·cm	h %	Note	Ref.
d(pT-T-G)			260	29.4			59
U-A-Gp	1	0.1 M HCl	259	34.0		m	4
	1	0.1 M HCl	260*	33.8		9.9 (231), 0.83, 0.43, 0.21^m	4
	1	0.1 M NaCl, HClO$_4$	259	34.0	5.7(260)		112
	7	0.05 M phosp.	258	32.9		m	4
	7	0.05 M phosp.	260*	32.6	11.7	8.8(228), 0.87, 0.41, 0.15^m	4 4
	7	0.1 M NaCl	258	32.9	11.7 (260)		112
	13	0.1 M KOH	260	31.9		14.7(233), 0.82, 0.39, 0.08^m	4
	13	0.1 M NaCl, NaOH	260 260	31.9	6.5 13.2		112 111
U-C-G	8.6		260*	27		257(262), 0.93(0.79), 0.60(0.81), 0.23(0.50)^n	284
U-G-A		0.2 N NaOH	260*		6.8"	a.h.^{c}	167, 169
U-U-G	8.6		260*	29		257(259), 0.92(0.86), 0.46(0.47), 0.14(0.20)^n	284
U-U-Gp	1	0.1 M HCl	261	28.9		m	4
	1	0.1 M HCl	260*	28.9		6.9 (230), 0.78, 0.50, 0.23^m	4
	1	0.1 M NaCl, HClO$_4$	261	28.9	8.2(260)		112
	7	0.05 M phosp.	259	29.0		m	4
	7	0.05 M phosp.	260*	28.9	7.8	8.2(229), 0.87, 0.48, 0.17^m	4
	7	0.1 M NaCl	259	29.0	7.8(260)		112
	13	0.1 M KOH	262	24.7		m	4

Oligomers	pH	Medium	λ mμ	$E \times 10^{-3}$ 1/mol·cm	h %	Note	Ref.
U-U-Gp (Con't)							
	13	0.1 M KOH	260*	24.6		16.7(238), 0.84,0.48, 0.10^m	4
	13	0.1 M NaCl,NaOH	262	24.7	6.5(260)		112
hU-Gm-G	8.6		260*	23		252(255), 1.10(0.96), 0.64(0.64), 0.32(0.42)^n	284
A-A-A-Cp	1		260*	48.9		$E_{min}(233)=$ 15.0	158
	1		259	49.0		0.76,0.46^r	158
	7		260*	39.9	25.3	$E_{min}(230.5)$ =13.8	158
	7		259	40.0		0.77, 0.52^r	158
	13		260*	39.6		$E_{min}(231)=$ 15.6	158
	13		259	39.7		0.78$_5$, 0.52$_5$^{r}	158
A-A-A-G	8.6		260*	44		257(257), 0.92(0.88), 034(0.34), 0.10(0.13)^n	284
A-A-A-Gp	1		260*	50.2		$E_{min}(229.5)=$ 12.6	158
	1		257	51.4		0.86,0.33^r	158
	7		260*	44.9	22.2	$E_{min}(228)=$ 11.6	158
	7		256.6	46.3		0.89,0.25^r	158
	13		260*	46.3			158
	13		258	47.0		0.84,0.336$_6$^{r}	158
A-A-A-U	8.6		260*	40.0		258(258), 0.79(0.81), 0.27(0.26), 0.02(0.04)^n	55

Oligomers	pH	Medium	λ $m\mu$	$E \times 10^{-3}$ $1/mol \cdot cm$	h %	Note	Ref.
A–A–A–Up	1		260*	<u>48.6</u>		$E_{min}(230)=$ <u>12.5</u>	158
	1		258	<u>48.8</u>		0.83,0.26[r]	158
	7		260*	<u>42.8</u>	23.7	$E_{min}(229.5)$ =<u>7.3</u>	158
	7		259	<u>43.0</u>		0.76,0.29[r]	158
	8.6		260*	<u>40.0</u>		258(258), 0.79(0.81), 0.27(0.28), 0.02(0.04)[n]	55
	13		260*	<u>41.7</u>			158
	13		258	<u>42.0</u>		0.835,0.25[r]	158
A–G–G–Tp	8.6		260*	<u>42</u>		256(256), 0.97(0.96), 0.51(0.52), 0.22(0.25)[n]	284
A–U–G–C		0.1 and 0.2 $\underline{N}$ NaOH	260*		8.6"	a.h.[c]	166, 167, 169
d(A–T)$_n$	(N)	VAR[Na$^+$]	262*		VAR	n = 9–22	217
C–C–C–G	8.6		260*	<u>30</u>		270(278) 0.94(0.67), 0.82(1.44), 0.36(1.00)[n]	284
C–G–A–U		0.2 $\underline{N}$ NaOH	260*		12.9"	a.h.[c]	167, 169
C–G–U–A		0.2 $\underline{N}$ NaOH	260*		6.2"	a.h.[c]	167, 169
C–U–G–A		0.2 $\underline{N}$ NaOH	260*		7.7"	a.h.[c]	167, 169
C–U–C–U–G	8.6		260*	<u>42</u>		260(265), 0.88(0.76), 0.59(0.84) 0.24(0.52)[n]	284

Oligomers	pH	Medium	λ mμ	E x 10^{-3} l/mol·cm	h %	Note	Ref.
C-U-U-U-G	8.6		260*	44		259(261), 0.86(0.78), 0.53(0.63), 0.17(0.32)^n	284
C-A-A-C-U-U-G	8.6		260*	66		260(261), 0.86(0.78), 0.53(0.63), 0.23(0.36)^n	284
G-A-A-A		H_2O	256				104
G-A-br-^{8}A-br^{8}A		H_2O	260				104
G-A-G-Up	8.6		260*	42		256(257), 0.96(0.90), 0.45(0.50), 0.18(0.27)^n	284
G-A-G-hUp	8.6		260*	35		255(256), 0.98(0.93), 0.48(0.50), 0.22(0.29)^n	284
G-C-A-U		0.2 N NaOH	260*		12.3"	a.h.^{c}	167, 169
G-C-U-A		0.2 N NaOH	260*		6.6"	a.h.^{c}	167, 169
T-Q-C-G	8.6		260*	32		260(265), 0.86(0.75), 0.64(0.84), 0.25(0.49)^n	284
U-C-C-U	8.6		260*	31.3		263,0.86, 0.63,0.25^n	55
U-C-C-U-G	8.6		260*	42		260(265) 0.88(0.75), 0.59(0.84), 0.24(0.52)^n	284
hU-hU-A-A-G	8.6		260*	35		256(257), 0.93(0.88), 0.38(0.38), 0.14(0.18)^n	284

Oligomers	pH	Medium	λ mμ	E x 10^{-3} 1/mol·cm	h %	Note	Ref.
Um–G–G–G–C	8.6		260*	<u>47</u>		257(256), 0.95(0.96), 0.63(0.68), 0.34(0.45)^n	284

Oligomers	pH	Medium	λ mμ	E x 10^{-3} 1/mol·cm	h %	Note	Ref.

2.1.3. Sugar-Phosphate Backbone Analogues

Oligomers	pH	Medium	λ mμ	E x 10^{-3} 1/mol·cm	h %	Note	Ref.
a(s^{8,2}As^{8,2}A)	1		276.5	42			105
	7		271	38			105
	13		271.5	35.8			105
A-U4ss4U5'-3'A	6.2		261	17.3		290(5.0)[p]	114
	6.2		318	16.8			114
G-U4ss4U5'-3'G	6.0		317	19.1		299(7.7)[p]	114
U4ss4U	6.0		261	12.3		280(10.0)	114
	6.0		310	19.7			114
Ade-C$_3$		0.1 N HCl	259.5	14.08		2.82(231)[t]	24
		H$_2$O	261	14.36		2.47(227)[t]	24
		0.1 N NaOH	261	14.34		2.38(232.5)[t]	24
Ade-C$_2$-Ade		0.1 N HCl	257	26.24		5.40(228)[t]	24
		H$_2$O	257	24.65		4.46(227.5)[t]	24
		0.1 N NaOH	257	24.65		4.44(227.5)[t]	24
Ade-C$_3$-Ade		0.1 N HCl	257	26.05		6.19(231)[t]	24
		H$_2$O	256	24.41		5.34(228.5)[t]	24
		0.1 N NaOH	256	24.25		5.22(229)[t]	24
Ade-C$_6$-Ade		0.1 N HCl	259.5	28.19		5.75(231)[t]	24
		H$_2$O	258	26.30		4.68(227.5)[t]	24
		0.1 N NaOH	258	26.31		4.58(228)[t]	24
Ade-C$_3$-Cyt		0.1 N HCl	267	19.47		5.27(234.5)[t]	24
			285sh	14.70			24
		H$_2$O	262.5	18.26		9.26(232.5)[t]	24
		0.1 N NaOH	262.5	18.29		9.34(232.5)[t]	24
Ade-C$_3$-Gua		0.1 N HCl	256	23.25		5.74(228)[t]	24

Oligomers	pH	Medium	λ mμ	$E \times 10^{-3}$ 1/mol·cm	h %	Note	Ref.
Ade-C$_3$-Gua (Con't)		H$_2$O	253	21.53		6.02(228)[t]	24
		0.1 N NaOH	259	21.35		6.85(230)[t]	24
Ade-C$_3$-Thy		0.1 N HCl	261.5	20.19		5.54(233)[t]	24
		H$_2$O	261.5	19.85		5.03(232)	24
		0.1 N NaOH	262	18.29		8.11(237.5)	24
Ade-C$_3$-Ura		0.1 N HCl	260.5	22.32		4.88(231.5)[t]	24
		H$_2$O	260.5	21.63		4.18(229.5)[t]	24
		0.1 N NaOH	261	20.02		7.65(234)[t]	24
Cyt-C$_3$		0.1 N HCl	214.5	9.78		0.98(241.5)[t]	24
			283.5	12.76			24
		H$_2$O	230sh	7.54		4.02(249)[t]	24
			274	8.60			24
		0.1 N NaOH	230sh	7.72		4.06(249)[t]	24
			274	8.53			24
Cyt-C$_3$-Cyt		0.1 N HCl	213	18.71		2.16(241)[t]	24
			283	25.56			24
			230sh	14.60		8.27(249)[t]	24
			274	15.91			24
		0.1 N NaOH	230sh	14.40		8.35(249.5)[t]	24
			273.5	15.80			24
Cyt-C$_3$-Gua		0.1 N HCl	256.5	14.15		5.84(232)[t]	24
			281	18.54		13.90(262.5)[t]	24
		H$_2$O	253.5	14.72		10.20(227.5)[t]	24
			270.5	14.29		13.80(262)[t]	24
		0.1 N NaOH	269	15.69		9.80(237.5)[t]	24

Oligomers	pH	Medium	λ mμ	E x 10^{-3} 1/mol·cm	h %	Note	Ref.
Cyt —C$_3$-Thy		0.1 N HCl	277	19.99		3.07(237.5)^t	24
		H$_2$O	215sh	20.40		7.12(242)^t	24
			271.5	16.49			24
		0.1 N NaOH	230sh	16.20		7.30(246.5)^t	24
			271	13.69			24
Gua-C$_3$		0.1 M HCl	252	11.86		2.57(225)^t	24
			277	7.76		7.34(268)^t	24
		H$_2$O	252	12.51		3.16(225)^t	24
			270sh	9.31			24
		0.1 N NaOH	255sh	9.90		4.42(231)^t	24
			267.5	10.58			24
Gua-C$_3$-Gua		0.1 N HCl	252.5	22.43		8.47(226.5)^t	24
			275sh	13.20			24
		H$_2$O	250	20.68		6.11(226.5)^t	24
			270	15.20			24
		0.1 N NaOH	257	17.80		8.54(231.5)^t	24
			267	18.48			24
Gua-C$_3$-Thy		0.1 N HCl	257.5	16.69		5.95(230.5)^t	24
			270	15.60			24
		H$_2$O	256	16.50		6.77(230.5)^t	24
			267	16.09		16.00(263.5)^t	24
		0.1 N NaOH	267	16.65		8.99(238)^t	24
Thy-C$_3$		0.1 N HCl	272.5	9.59		1.68(237)^t	24
		H$_2$O	272	9.61		1.63(237)^t	24
		0.1 N NaOH	270	7.20		4.29(244)^t	24
Thy-C$_3$-Thy		0.1 N HCl	269.5	17.38		3.57(236)^t	24

Oligomers	pH	Medium	λ $m\mu$	E x 10^{-3} 1/mol·cm	h %	Note	Ref.
Thy-C$_3$-Thy (Con't)							
		H$_2$O	269	17.36		3.53(236)[t]	24
		0.1 N NaOH	269.5	13.28		6.45(244)[t]	24
Ura-C$_3$		0.1 N HCl	267	10.13		1.33(231.5)[t]	24
		H$_2$O	267	10.13		1.37(232)[t]	24
		0.1 N NaOH	265	7.35		3.48(241)[t]	24
Ura-C$_3$-Ura		0.1 N HCl	266.5	19.73		2.94(231.5)[t]	24
		H$_2$O	266.5	19.73		2.97(232)[t]	24
		0.1 N NaOH	265.5	14.50		7.24(241.5)[t]	24

FOOTNOTES FOR SECTION 2.1

a The values are λ_{min} and absorbance ratios relative to 260 mμ
   at 250, 280, and 290 mμ, respectively. The following is applicable only for ref. 188: λ_{max} and λ_{min} are accurate within
   $\pm$1 mμ and E_{max} (related to P = 31) within $\pm$3% .

b Calculated from the hypochromicity, assuming E_{max} (259) of
   adenosine and adenosine-2'(3')-monophosphate to be 15.4 x 10^{-3}.

c See the ref. 166 or 167, whichever is applicable, for λ_{max},
   λ_{min}, and A_{280}/A_{260} values in 0.01 $\underline{N}$ HCl, 0.01 $\underline{N}$ NaOH, and
   0.1 $\underline{N}$ NaOH.

d Determined by the same method, E_{260} (AMP) = (14.6 $\pm$ 0.07) x 10^3.

e E values for dinucleoside monophosphates are based on literature
   values of E x 10^{-3} of dCMP, 7.4 (at 260 mμ); dTMP, 9.55 (266);
   dAMP, 15.4 (260); dGMP, 11.5 (260); the latter two values were
   assumed to be the same as those for ribonucleotides. The hypochromicity at 260 mμ of a deoxydinucleoside phosphate was assumed to be the same as that of the corresponding ribo-com-
   pound. Terminal phosphates were neglected.

f E_{max} values are at 21-23°. The first value in the Note column is
   extinction coefficient obtained from the least squares line of
   a plot of the observed E_{max} vs. 1/n, where n is the chain
   length of the oligonucleotide; the second value, when given,
   is E_{max} at 90°. E_{max} x 10^{-3} is 15.3 at 21-23° and 14.85 at 90°
   for dAMP (259 mμ) and 9.65 at 21-23° for dTMP (267 mμ).

g The first value is λ_{max}; the second value, when given, is
   E_{260} x 10^{-3} calculated from the equation $E_{IJK}(\lambda) = (1/3) \times$
   $[2E_{LJ}(\lambda) + 2 E_{JK}(\lambda) - E_{J}(\lambda)]$, where E_{IJK} is the extinction
   coefficient per residue of the trinucleoside diphosphate IpJpK.
   The other terms are defined analogously. The experimental
   values were determined upon hydrolysis of IpJpK by Worthington
   venom phosphodiesterase.

h Values in parentheses are those obtained by nearest neighbor
   calculations.

i See the reference for spectral data of arabinosylcytosine monophosphates and for approximate molar absorbancies at λ_{max} (pH 2
   and 5) of various dinucleoside monophosphates containing
   arabinosylcytosine and 2'-5', 3'-5', and 5'-5' linkages.

j Based on E_{266} of TMP = 9,550; spectra were measured and hydrolysis by snake venom phosphodiesterase was carried out at 25°C.

k See the reference for data of some acetylated and phosphorylated
   derivatives.

l The values are λ_{max}, λ_{min}, and absorbance ratios relative to
   260 mμ at 250, 280, and 290 mμ, respectively.

m Molar extinction coefficients for all pH values were calculated
   from hypochromicities at 260 mμ upon alkaline hydrolysis using
   the following values of E_{260} x 10^{-3} (pH 7): 15.4, 7.3, 11.6,
   and 9.9, for 2'(3')-monophosphates of adenosine, cytidine,
   guanosine, and uridine, respectively. All absorbance measurements were made at 20°C. The values in the Note column, when
   given, are E_{min} (λ_{min}) and absorbance ratios relative to
   260 mμ at 250, 280, and 290 mμ, respectively.

n E values are similar to those reported in refs. 51, 163, and 164,
   but were reestimated considering other pertinent data. The
   data in the Note column, which were obtained in water and
   0.02 $\underline{N}$ HCl (in parentheses), are λ_{max} and absorbance ratios
   relative to 260 mμ at 250, 280, and 290 mμ, respectively.

o Same as in note n, except the values in parentheses were
   measured in 0.2 $\underline{N}$ NH$_4$OH.

p E values, determined by hydrolysing the dimer with RNase T_2, are
   based on E_{260} x 10^{-3} of AMP, GMP, and 4-thiouridine being 15.4,
   11.6, and 3.3, respectively, and E_{330} of 4-thiouridine being
   20.6 x 10^3 (pH 5.8, I = 0.1 $\underline{M}$). The values in the Note column
   are λ_{min} (E_{min} x 10^{-3}).

r A_{250}/A_{260} and A_{280}/A_{260}.

s 4-thiouridine (pH 6.0): λ_{max} (E_{max} x 10^{-3}) = 245 (4.6), 330
   (20.6); and λ_{min} (E_{min} x 10^{-3}) = 280 (2.3).

t E_{min} x 10^{-3} (λ_{min}).

[u] The λ_{max} (mμ), E_{max} x 10^3, hyperchromicity (%), and λ_{min} (mμ) values for bis[6-dimethylamino-9-β-D-(3'-acetamido-3'-deoxy)-ribofuranosyl-5'-]phosphate are as follows: (pH 2) 267, 32.6, 11.9, 235;   (pH 7) 274, 27.8, 33.1, 235;   (pH 13) 274, 27.0, 39.3, 235.

2.2   COMPLEXES OF POLYNUCLEOTIDES WITH THEIR COMPONENTS

Complex	T °C	pH	Medium	λ mμ	E x 10^{-3} 1/mol·cm	Note	Ref.
Poly A . d(Tp)$_n$	0.5	6.9	0.4 $\underline{M}$ NaCl			H $\underline{vs.}$ n	35
	0.5	6.9	1 $\underline{M}$ LiCl			H $\underline{vs.}$ n	35
Poly U . n^{2}m^{6}Ado		7.5	0.54 Na$^+$, 0.13 $\underline{M}$ phosp.	263.5	8.9	E_{min}(241)=4.97	102
				~282sh	~5.8		

2.3.   POLYNUCLEOTIDES

Polymer	T $^\circ$C	pH	Medium	λ mμ	$E \times 10^{-3}$ 1/mol·cm	Note	Ref.

2.3.1.   Homopolymers

Polymer	T $^\circ$C	pH	Medium	λ mμ	$E \times 10^{-3}$ 1/mol·cm	Note	Ref.
Poly A	10	4.00	0.1 M KCl	259*	7.60	B,100% hx.[a]	206
	25	4.00	0.1 M KCl	259*	7.60	B,100% hx.[a]	206
	22	4.00	0.1 M citr.	252	8.7	Acid A form[d]	2
	22	5.81	0.1 M citr.	253	8.8	Acid B form[d]	2
	22	8.5	0.1 M citr.	257	10.1	Neutral form[d]	2
	22	3.76[b]	0.1 M citr.;60% diox.	259	9.1	Acid form[d]	2
	22	8.5	0.1 M citr.;50% diox.	259	13.0[c]	Neutral form[d]	2
	25	4.2	0.1 M NaCl	252	(9.5)		125
		4.5	0.15 M acet.	251			198
		4.5	0.1 M NaCl,0.05 M acet.	252	8.6	B;a.h.	20
	25	4.6	0.15 M KCl	252	8.2	B;d.s.hx;40%H[e]	92
	83	4.6	0.15 M KCl	258	12.0	B;s.s.hx;21%H[e]	92
	22	5.4	0.15 M KCl	252	8.2	B;d.s.hx;40%H[e]	92
	48	5.4	0.15 M KCl	256	10.2	B;s.s.hx;32%H[e]	92
	82	5.4.	0.15 M KCl	258	12.2	B;s.s.hx;21%H[e]	92
	22	7.1	0.15 M KCl	256	10.1	B;s.s.hx;26%H[e]	92
	44	7.1	0.15 M KCl	257	11.3	B;s.s.hx;25%H[e]	92
	26	4.85	0.1 M NaCl,0.1 M Na acet.	(250)	8.6	100% hx.	85
	78	4.85	0.1 M NaCl,0.1 M Na acet.	(259)	11.9	43% hx.	85
	26	4.85	0.1 M NaCl,0.1 M Na acet.	252	8.6	P	269
	78	4.85	0.1 M NaCl,0.1 M Na acet.	257	11.9	P	269
	VAR	4.85	0.1 M NaCl,0.1 M Na acet.	max.	VAR		269

Footnotes for Section 2.3 start on page 95.

Polymer	T °C	pH	Medium	λ mμ	E x 10^{-3} 1/mol·cm	Note	Ref.
Poly A (Con't)							
	20	4.9	0.1 M NaCl,0.1 M acet.	252	8.6	B	275
		5.5	0.05 M KF,1 mM EDTA	251	8.60	257,15.4[f]	70
			0.05 M KF,1 mM EDTA;e.g.	258	12.80	258,14.62[f]	70
	4	6.0	0.1 M phosp.,1 mM Na cacd.,1 mM EDTA	(251.5)			87
	23.1	6.0		(253.5)			87
	34.3	6.0		(258)			87
	25	6.4	0.1 M NaCl	257	(10.6)		125
	25	6.8	phosp.,I = 0.05 M	256	9.9±0.1	a h.;hpo.47%; 0.313[g]	118
	25	6.8	phosp.,I = 0.05 M	260*	9.5±0.1	a.h.;hpo.58%	118
	25	6.8	phosp.,I = 0.2 M	256	9.1±0.2	a.h.;hpo.52%; 0.319[g]	118
	25	6.8	phosp.,I = 0.2 M	260*	9.6±0.2	a.h.;hpo.64%	118
		6.8-7.0	0.01 M cacd.	255*	10.6	E_{259}(AMP)=15.1; a.h.	129
		6.8-7.0	0.01 M cacd.	257*	10.5	B;a.h.	129
		6.8-7.0	0.01 M cacd.	VAR	VAR	a.h.	129
		6.85	0.1 M NaCl,0.01 M Na cacd.	257*	10.0±0.2	P;p.a.	228
		6.95	0.1 M NaCl,0.01 M Na cacd., 1 mM MgCl$_2$		10.4	B	228
		7.0	0.01 M K phosp.	256.5	10.7±0.4	P;p.a.; 0.97,0.30[h]	233
		7.0	0.01 M phosp.	257	10.7	P	21
		7.0	0.0125 M Na cacd.	256.5		0.67,0.32[i]	104
	20	7.0	0.1 M NaCl,0.1 M Tris	256.5	9.1	B	275
	85	7.0	0.1 M NaCl,0.1 M Tris	257.5	11.4	B	275
	10	7.0	0.1 M KCl	259*	9.44	B;58.5% hx.[d]	206
	25	7.00	0.1 M KCl	259*	10.10	B;43.5% hx.[d]	206

Polymer	T °C	pH	Medium	λ mμ	$E \times 10^{-3}$ 1/mol·cm	Note	Ref.
Poly A (Con't)							
	40	7.00	0.1 M KCl	259*	10.85	B;26.5%hx.[d]	206
	80	7.00	0.1 M KCl	259*	12.02	B;0.0%hx.[d]	206
		7	0.01 M K phosp.	256.5	10.7± 0.4	0.97;0.30[h]	233
		7	0.01 M LiCl,0.01 M cacd.	259*	10.1		135
		7	phosp.	257	9.8	P;a.h.	278
		7		257	10.5	B;a.h.	130
	83	7.1	0.15 M KCl	258	12.9	B;s.s.hx;14%H[e]	92
		7.4	0.1 M NaCl,0.01 M Tris	256.5	9.0	B;a.h.	20
	26	7.5	0.1 M NaCl,0.05 M Tris	257	10.5	P;259,15.4[f]	269
	80	7.5	0.1 M NaCl,0.05 M Tris	257	12.5	P;259,15.4[f]	269
	26	7.5	0.1 M NaCl,0.05 M Tris, 4 mM MgCl$_2$	257	9.9	P	269
	VAR	7.5	0.1 M NaCl,0.05 M Tris	max.	VAR		269
	26	7.5	0.1 M NaCl,0.05 M Tris	(256)	10.5	67% hx.	85
	80	7.5	0.1 M NaCl,0.05 M Tris		12.5	33% hx.	85
	26		formamide	(259)	13.8	0% hx.	85
	26	7.5	0.1 M NaCl,0.05 M Tris, 4 mM MgCl$_2$		9.9	78% hx.	85
	25	7.75-7.80	0.1 M Na$^+$	257	10.7	P;a.h.; 260,15.0[f]	210
	25	7.75-7.80	0.1 M Na$^+$	260*	10.3	P;a.h.	210
	25	7.75-7.80	0.1 M Na$^+$,0.01 M Na phosp.	257	10.7	P;p.a.	37
				257	9.9	P;a.h.;p.a.	214
		8.0	1 mM Tris		10.3	(B)	286
	27 ±1	8.8	0.05 M KF, 1 mM Tris	257	10.54	258,15.4[f]	70
	27±1		0.05 M KF,1 mM Tris;e.g.	259	15.08	258,14.3[f]	70
	26		formamide	260	13.8	P;261,14.8[f]	269
Poly m^{1}A		4.5	0.15 M Na acet.	255			180
Poly m^{6}A		4.5	0.15 M Na acet.	258			180

Polymer	T $^{\circ}$C	pH	Medium	λ mμ	E x 10^{-3} 1/mol·cm	Note	Ref.
Poly he^{6}A	20	4.9	0.1 $\underline{M}$ NaCl,0.1 $\underline{M}$ Tris	265	11.5	B;a.h.;s.v.h.	275
	20	7.0	0.1 $\underline{M}$ NaCl,0.1 $\underline{M}$ Tris	265	11.5	B;a.h.;s.v.h.	275
	85	7.0	0.1 $\underline{M}$ NaCl,0.1 $\underline{M}$ Tris	266	13.0	B;a.h.;s.v.h.	275
Poly n^{2}A		7.9	0.1 $\underline{M}$ NaCl,0.01 $\underline{M}$ pyro-phosp.	258; 279	6.71; 6.03	255,9.45^f 280,10.0	96
Poly n^{2}m^{6}A	10.8	4.7	0.025 $\underline{M}$ acet.	256;291 ~310sh		239,275min	102
	10.7	5.3	0.025 $\underline{M}$ acet.	256.5;287 ~308sh		240,276min	102
		7.5	0.1 $\underline{M}$ Na$^+$, 0.02 $\underline{M}$ phosp.	265;277 ~295sh	8.7;9.2 ~4.5	E_{min}(243)=4.4^j	102
Poly c^{7}A		7.0	0.05 $\underline{M}$ Na cacd.	270	(7.5)	hpo.25%	103
Poly Am	rm	7	0.15 $\underline{M}$ KCl,0.01 $\underline{M}$ cacd.	257	9.9	ac.h.;14-15S	12
Poly Aac	25	4.05	0.1 $\underline{M}$ NaCl	257	(10.7)	88% acetylation	125
	25	7.3	0.1 $\underline{M}$ NaCl	257	(12.5)	88% acetylation	125
Poly dA	22	3.02	0.1 $\underline{M}$ citr.	261	11.5	Acid A form^d	2
	22	4.30	0.1 $\underline{M}$ citr.	261	11.5	Acid B form^d	2
	22	8.5	0.1 $\underline{M}$ citr.	257.5	9.9	Neutral form^d	2
	22	3.16^b	0.1 $\underline{M}$ citr.;60% diox.	259	11.0	Acid form^d	2
	22	8.5	0.1 $\underline{M}$ citr.;60% diox.	263	12.7	Neutral form^d	2
	25	7.75-7.80	0.1 $\underline{M}$ Na$^+$,0.01 $\underline{M}$ phosp.	257	8.6	P; a.h.,p.a.	37,210
	25	7.75-7.80	0.1 $\underline{M}$ Na$^+$,0.01 $\underline{M}$ phosp.	260	8.4	P; p.a.; 260,15.0^f	210
		8.0	1 m$\underline{M}$ Tris	257	10$\pm$0.2	B,P; ac.h.	271
		8.0	0.1 $\underline{M}$ NaCl,1 m$\underline{M}$ Tris	max	9.8$\pm$0.1	B,P; ac.h.	271
		8.0	VAR [Na$^+$]	VAR	VAR		271
		8.0	1 m$\underline{M}$ Tris	260*	9.65	P;0.91,0.29^h	14

Polymer	T °C	pH	Medium	λ $m\mu$	$E \times 10^{-3}$ $1/mol \cdot cm$	Note	Ref.
Poly dA (Con't)							
		8.0	1 mM Tris		9.65	(B),S_{20}W= 5.04(pH 7)	286
	21-23	8.0	1 mM Tris	257	9.39	n > 300	34
	90	8.0	1 mM Tris	max	12.04	n > 300	34
	VAR		VAR salt concn.	VAR	VAR		268
Poly C	25	3.65	0.1 M salt	277	8.0	P; a.h.[s]	3
	25	4.05	0.1 M salt	274	7.7	P; a.h.	3
	25	4.85	0.1 M salt	274	7.2	P; a.h.	3
	25	7.5	0.1 M salt	267	6.6	P; a.h.	3
	26	4.0	0.1 M Na acet.	275	8.9	P;242,3.0[k] 276,11.1[f]	269
	84.5	4.0	0.1 M Na acet.	275	7.4	P;243,3.3[k]; 274,9.3[f]	269
	VAR	4.0	0.1 M Na acet.	max	VAR		269
		4.0	0.1 M NaCl,0.05 M acet.	275	6.5		18
	20	4	0.15 M Na$^+$	(277)	(7.8)		175
	26	4.85	0.1 M NaCl,0.1 M Na acet.	274	7.0	P;244,3.3[k]; 273,9.5[f]	269
	80	4.85	0.1 M NaCl,0.1 M Na acet.	270	7.7	P;249,5.7[k]; 271,9.0[f]	269
	VAR	4.85	0.1 M NaCl,0.1 M Na acet.	max	VAR		269
		5.5	0.5 M KF,1 mM EDTA	273	7.4	270,9.0[f]	70
			0.05 M KF,1 mM EDTA;e.g.	272	8.12	272,8.3[f]	70
		6.85	0.1 M NaCl,0.01 M Na cacd.	258*	6.3±0.1	P; p.a.	228
		7.0	phosp.	269	6.3	P; a.h.	278
		7.0	0.01 M K phosp.	268	6.8	P	21
	20	7	0.15 M Na$^+$	(268)	(7.5)		175
		(N)	0.1 M salt	268*	6.3	P; a.h.,p.a.	214
		7.5	0.1 M KF,0.01 M Tris	268	6.09		18
	26	7.5	0.1 M NaCl,0.05 M Tris	267	6.5	P;248,4.9[k]; 271,9.0[f]	269
	80	7.5	0.1 M NaCl,0.05 M Tris	269	7.4	P;252,6.1[k]; 271,8.7[f]	269

Polymer	T $^\circ$C	pH	Medium	λ mμ	E x 10^{-3} 1/mol·cm	Note	Ref.
Poly C (Con't)							
	VAR	7.5	0.1 $\underline{M}$ NaCl,0.05 $\underline{M}$ Tris	max	VAR		269
	25	7.6	phosp.; I = 0.05 $\underline{M}$	268	6.5$\pm$0.1	p.r.h.;0.785^g; hpo.41%	118
	25	7.6	phosp.; I = 0.05 $\underline{M}$	260*	5.8$\pm$0.1	p.r.h.;hpo.33%	118
		7.7	0.01 $\underline{M}$ Tris	270*		hpo.35%[1]	120
		7.75–7.80	0.1 $\underline{M}$ Na$^+$,0.01 $\underline{M}$ Na phosp.	268	6.2	P; p.a.	37,40
	VAR	7.8	0.1 $\underline{M}$ NaCl,0.05 $\underline{M}$ Tris	270	VAR	a.h.,w.p.h.	257
		8.8	0.05 $\underline{M}$ KF, 1 m$\underline{M}$ Tris	268	6.5	270,9.0^f	70
			0.05 $\underline{M}$ KF, 1 m$\underline{M}$ Tris;e.g.	273	8.24	273,8.28^f	70
Poly m^{5}C	25	4.0	0.1 $\underline{M}$ buff.	280	6.7	hper.58.5%; 284,10.7^f	249
	85	4.0	0.1 $\underline{M}$ buff.	281.5	8.65	hper.5.2%; 282,8.2^f	249
	25	7.8	0.1 $\underline{M}$ buff.	275	5.8	hper.52%; 278.5,8.8^f	249
	85	7.8	0.1 $\underline{M}$ buff.	277	6.65	hper.29%; 278.5,8.6^f	249
	VAR	7.8	0.1 $\underline{M}$ NaCl,0.05 $\underline{M}$ Tris	277	VAR	a.h.,w.p.h.	257
Poly br^{5}C	20	4	0.15 $\underline{M}$ Na$^+$	292	(4.2)		175
	80	4	0.15 $\underline{M}$ Na$^+$	(290)	(4.6)		175
	20	7	0.15 $\underline{M}$ Na$^+$	(289)	(4.2)		175
	25	7.0	0.2 $\underline{M}$ Na cacd.	289	5.51	P;265,3.25^k	97
	93	7.0	0.2 $\underline{M}$ Na cacd.	289	6.14	P	97
Poly io^{5}C	20	4	0.15 $\underline{M}$ NaCl,0.05 $\underline{M}$ Na acet.	303	(3.7)		175
	60	4	0.15 $\underline{M}$ NaCl,0.05 $\underline{M}$ Na acet.	(302)	(3.9)		175
	85	4	0.15 $\underline{M}$ NaCl,0.05 $\underline{M}$ Na acet.	(298)	(4.3)		175
	20	7	0.15 $\underline{M}$ Na$^+$	(298)	(3.8)		175
Poly ho^{4}C		7		235;(268)		hpo.12%(260), 13%(235)	234
		7.7	0.01 $\underline{M}$ Tris	270*		hper.2.0%[1]	74

Polymer	$T_{°C}$	pH	Medium	λ $m\mu$	$E \times 10^{-3}$ 1/mol·cm	Note	Ref.
Poly mh^{4}C		7		(233.5;272)		hpo.9%(260), 11%(235)	234
Poly Cm		N		268		e.h.; hper.37%	121
Poly dC		6.4		274	6.6	P;p.a.;h.p.hx.^{m}	106
		N		max.		hper.18%	287
		7.75-7.80	0.1 M Na$^+$,0.01 M Na phosp.	270	7.2	P; p.a.	37,40
		8.0	1 mM Tris	260*	5.30	P;0.77,1.27^h	14,287
		8.0	1 mM Tris		5.3)	(B);$S_{20,w}=$ 4.80(pH 8.0)	286
		10.0		269	7.5$_5$	P; p.a.^{m}	106
Poly d(m^{5}C)		N		max		hper.21%	287
		8.95	0.03 M Tris,0.01 M MgCl$_2$	260*	4.70	P; p.d.h.; 1.03,1.33^h	287
Poly d(m^{4,5}C)		N		max		hper.28%	287
		8.95	0.03 M Tris,0.01 M MgCl$_2$	260*	6.47	P; p.d.h.; 0.96,1.08^h	287
Poly d(m^{5}e^{4}C)		N		max		hper.16%	287
		8.95	0.03 M Tris,0.01 M MgCl$_2$	260*	7.71	P; p.d.h.; 0.98,1.13^h	287
Poly d(br^{5}C)		3.7		293	5.2	P;h.p.hx.^{m}	106
		7.0		288	5.5	P; n.p.c.	106
Poly G		5.5	0.05 M KF, 1 mM EDTA	253	9.5	250,13.7^f	70
			0.05 M KF, 1 mM EDTA;e.g.	255	9.35	255,15.1^f	70
	25	7.4	0.6 M Na$^+$	(260)	(9.7)		62
	25	12.5	0.6 M Na$^+$	(255)	(11.1)		62

Polymer	T $^\circ$C	pH	Medium	λ mμ	E x 10^{-3} 1/mol·cm	Note	Ref.
Poly m^{2}G		2		263			198
		7		258			198
		11		263			198
Poly m^{7}G		2		260		30%G; 278-85sh; 233min^n	86
		7.4		260		30%G; 278-85sh; 233 min^n	
		12		269		30%G; 243min^n	86
Poly m^{2,2}G		2		264			198
		7.0	0.15 $\underline{M}$ NaCl	256			198
		11		262			198
Poly m^{2,2,7}G		5.0		262;290			
		9.3		230;258;287			198
Poly br^{8}G	20	7.0	0.1 $\underline{M}$ NaCl,0.05 $\underline{M}$ Na cacd.	(260)	(11.8)	<5% G	179
	90	7.0	0.1 $\underline{M}$ NaCl,0.05 $\underline{M}$ Na cacd.	(260)	(12.7)	<5% G	179
Poly dG		8.0	1 m$\underline{M}$ Tris	260*	8.99	P;1.15,0.54^h	14
Poly I		6.8	phosp.; I = 0.2 $\underline{M}$	247	7.9$\pm$0.1	hpo.54%;0.338^g	118
				260*	5.4$\pm$0.1	hpo.33%	118
		6.85-6.90	0.1 $\underline{M}$ NaCl,0.01 $\underline{M}$ Na cacd.	248*	10.4$\pm$0.2	P; p.a.	228
		7	phosp.	247	7.6	P; a.h.	278
		7.75-7.80	0.1 $\underline{M}$ Na$^+$,0.01 $\underline{M}$ Na phosp.	248	10.0	P	37,40
			0.1 $\underline{M}$ salt	248*	10.2	P; a.h.,p.a.	271
Poly dI		7.75-7.80	0.1 $\underline{M}$ NaCl,0.01 $\underline{M}$ Na phosp.	246	9.4	P; p.a.	37,40
		8.0	1 m$\underline{M}$ Tris	260*	5.35	P; 1.64,0.36^h	14,287
Poly T		6.8	0.1 $\underline{M}$ NaCl,0.01 $\underline{M}$ Na cacd.	266	6.8	P	72

Polymer	T °C	pH	Medium	λ mμ	E x 10^{-3} 1/mol·cm	Note	Ref.
Poly T (Con't)							
		7	SSC	266		e.h.; hper.13%(37°); hper.(hx-c)32%	218
Poly m^{3}T		7		266		e.h.; hper.16%(37°)	218
Poly dT	25	7.75-7.80	0.1 M Na$^+$	265	8.7	P; p.a;$_f$ 268,9.5^f	37,210
		8.0	1 mM Tris	260*	8.14	0.67,0.62^h	14
	21-23	8.0	1 mM Tris	265	8.54	n > 300	34
Poly U			0.01 M HCl	260	8.5	P; s.v.h.; hper.16%	172
	20	4.85	0.1 M NaCl,0.1 M Na acet.	260	9.1	P; 262,10.0^f	269
	80	4.85	0.1 M NaCl,0.1 M Na acet.	260	9.2	P; 262,10.0^f	269
		5.5	0.05 M KF, 1 mM EDTA	261	9.2	263,10.0^f	70
	25		phosp.; I = 0.2 M	260	9.1$\pm$0.1	a.h.; hpo.9%; 0.354^g	118
		6.8-7.0	0.01 M cacd.	260	9.58	B; a.h.; E$_{261}$(UMP)=10.1	129
		6.8-7.0	0.01 M cacd.	261*	9.5	B; a.h.; E$_{261}$(UMP)=10.1	129
		6.8-7.0	0.01 M cacd.	VAR	VAR		129
		6.95	0.1 M NaCl,0.01 M Na cacd., 1 mM MgCl$_2$	max	9.9	B	243
		7.0	0.01 M K phosp.	260	9.4	P	21
	20	7.0	0.1 M NaCl,0.02 M Tris	260*	9.10$\pm$0.05	B; a.h.	231
	20	7.0	2 mM Tris, VAR [Na$^+$]	260*	VAR		230
	20	7	0.0 M Na$^+$	max	9.52	P; 10.1^f	174
	20	7	0.03 M Na$^+$	max	9.33	P	174
	20	7	0.3 M Na$^+$	max	9.04	P	174
	20	7	1.0 M Na$^+$	max	8.76	P	174
	2	7	0.15 M Na$^+$,0.01 M Mg^{++}	max	6.8	P	174
		7	phosp.	261	9.6	P; a.h.	278
		7	Na cacd.	261*	8.5	P	157

Polymer	T °C	pH	Medium	λ mμ	E x 10⁻³ 1/mol·cm	Note	Ref.
Poly U (Con't)							
	20	7	MeOH	max	8.70	P	174
		7		261	9.5	B; a.h.	130
		7	SSC	261		e.h.; hper.7.5%(37°); hper.(hx-c)18.5%	218
		(N)	0.05 $\underline{M}$ KF,1 m$\underline{M}$ EDTA; e.g.	261	9.75	261,9.8^f	70
	20-80	7.5	0.1 $\underline{M}$ NaCl,0.05 $\underline{M}$ Tris	260	9.2	P;262,10.0^f	269
	25	7.75-7.80	0.1 $\underline{M}$ Na$^+$	260	9.2	P;a.h.,p$_t$a.; 262,10.1^f	37,243
		7.8	0.01 $\underline{M}$ Tris			hper.9%^{1}	74
	20	11.6		260*	7.06±0.06	B;a.h.; E (UMP)=10.0	231
			0.01 $\underline{M}$ NaOH	260	6.4	P; s.v.h.	172
Poly m³U		7	Na cacd.	260		260^f	157
Poly m³br⁵U		7		279		279^f	157
Poly e⁵U	0	N	0.01 $\underline{M}$ Mg^{++}	(267)	(8.1)		247
	20	N	0.01 $\underline{M}$ Mg^{++}	(267)	(9.3)		247
Poly hm⁵U		7		263		e.h.; hper.7.0%(37°)	218
Poly fl⁵U	rm.		1 $\underline{N}$ HCl	269*		ac.h.; hper.8.5%	256
	rm.	6		269*		s.v.h.; hper.7.3%	256
		7	Na cacd.	267	6.5	P; 267^f	157
Poly cl⁵U		7	Na cacd.	275	7.5	P; s.v.h.; 275^f	157
			0.01 $\underline{M}$ HCl	275	7.0	P; s.v.h.; hper.26%	172
			0.01 $\underline{M}$ NaOH	272	5.2	P; s.v.h.	172
Poly br⁵U			0.01 $\underline{M}$ HCl	278	6.9	P; s.v.h.; hper.14.5%	172

Polymer	T °C	pH	Medium	λ mμ	$E \times 10^{-3}$ 1/mol·cm	Note	Ref.
Poly br^{5}U (Con't)		7		279	8.2	S_{20}(0.1 M NaCl)=8.1	211
		7	Na cacd.	279	7.4	P; 279^f	157
		12		277	6.1		211
			0.01 M NaOH	275	5.0	P; s.v.h.	172
Poly io^{5}U			0.01 M HCl	288	5.2	P; s.v.h.; hper.36.5%	172
		7	Na cacd.	289	5.1	P; 289^f	157
			0.01 M NaOH	281	3.7	P; s.v.h.	172
Poly ho^{5}U		7	Na cacd.	282	5	P; 282^f	157
Poly s^{4}U		7.0	0.05 M Na cacd.	(244;328)			232
Poly ms^{4}U		8.0	0.05 M Tris	(301)		320sh; A_{300}/A_{312}=1.40; hpo.40%(312)	220
Poly s^{2,4}U		7.0	0.05 M Na cacd.	(269)	(19.1)	t	53b
		7.0	0.05 M Na cacd.	260*	(15.2)	t	53b
		7.0	0.05 M Na cacd.	VAR		e.h.;VAR hper.	53b
Poly Um		9.0	0.1 M Tris,0.015 M MgCl$_2$	260		e.h.;hper.10%	289
Poly Q		7	Na cacd.	266	6.2	P; 266^f	157
Poly X		3.5		(237;263)	(6.6;5.7)		176
		7		(250)	(8.1)	269sh	176
		8.5		(248)	(9.2)		176
	25	6.8	phosp.; I = 0.2 M	250.5	8.1	a.h.;hpo.27%	118
	25	6.8	phosp.; I = 0.2 M	260*	7.2	a.h.;hpo.21%	118
Poly dX		8.0	1 mM Tris	260*	6.99	P; 1.08,0.61^h	14

Polymer	T °C	pH	Medium	λ mμ	E x 10^{-3} 1/mol·cm	Note	Ref.

2.3.2. Copolymers

Polymer	T °C	pH	Medium	λ mμ	E x 10^{-3} 1/mol·cm	Note	Ref.
Poly (A,br^{8}A)		7.0	0.0125 $\underline{M}$ Na cacd.	258		A/brA=15.6; 0.71,0.32^i	104
Poly (A,G)	25	6.7	0.1 $\underline{M}$ NaCl,1 m$\underline{M}$ EDTA, phosp.; I=0.15$\overline{3}$ $\underline{M}$	256.5	10.1	A/G=1.5; a.h.; hpo.40%;0.389^g	118
	25	6.7	0.1 $\underline{M}$ NaCl,1 m$\underline{M}$ EDTA, phosp.; I=0.15$\overline{3}$ $\underline{M}$	260*	9.6	A/G=1.5; a.h.; hpo.44%	118
		7.0	0.01 $\underline{M}$ K phosp.	255	11.2	P; A/G = 3.1	21
		7.0	0.01 $\underline{M}$ K phosp.	255	12.6	P; A/G = 4.5	21
Poly (A,br^{8}G)		7.0	0.0125 $\underline{M}$ Na cacd.	257		A/brG = 4.56; 0.70,0.35^i	104
		7.0	0.0125 $\underline{M}$ Na cacd.	257.5		A/brG = 3.36; 0.72,0.37^i	104
Poly (A,ho^{8}G)		7.0	0.0125 $\underline{M}$ Na cacd.	256.5		A/hoG=32.3; 0.66,0.31^i	104
Poly (A-T)		7.9	0.01 $\underline{M}$ Tris	264	5.7	P	277
Poly (A-s^{2}T)	rm		SSC	(269)		295sh	141
	12.2		NaOH	(244)		(260)sh	141
Poly d(A-T)		7.0	1 m$\underline{M}$ MgSO$_4$		6.6	P; p.a.	6
		N		260	9.6	P; e.h.; c.	110
		N		262	6.65	P; e.h.; hx.; hper.(hx-c)42±2%	110
	$\gg$ T$_m$	N		260	9.60±0.25	P; p.a.; c. 0.37^g	109
	$\gg$ T$_m$	N		262*	9.45	P; p.a.; c.	109
	$\ll$ T$_m$	N		262	6.65±0.05	P; p.a.; hx.; 0.55^g	109
Poly (A-U)	25	7.5	0.1 $\underline{M}$ Na citr.	260	6.1	P; a.h.; hx.; 261max;hpo.51%	38
	56	7.5	0.01 $\underline{M}$ Na citr.	260	10.4	P; a.h.; c.; 259max; hpo.16%	38

Polymer	T ^{o}C	pH	Medium	λ mμ	E x 10^{-3} 1/mol$\cdot$cm	Note	Ref.
Poly (A-U) (Con't)		7.9	0.01 $\underline{M}$ Tris	261	6.3	P	277
Poly (A-fl^{5}U)		7.9	0.01 $\underline{M}$ Tris	265	5.6	P	277
Poly (A-br^{5}U)		7.9	0.01 $\underline{M}$ Tris	268	5.5	P	277
Poly d(A-br^{5}U)	N $\ll T_m$			263	8.50$\pm$0.25	P; p.a.; c.; 0.57^g	109
	$\ll T_m$	N		266*	8.35	P; p.a.; c.	109
	$\gg T_m$	N		266	5.95$\pm$0.05	P; p.a.; hx.; 0.68^g	109
Poly (ho^{8}A,G)		7.0	0.0125 $\underline{M}$ Na cacd.	253.5		G/hoA = 24; 0.55,0.15^P	104
Poly (A,ms^{4}U)		8.0	0.05 $\underline{M}$ Tris	(260;304)		msU/A=1.25	220
Poly (A-Q)		7.9	0.01 $\underline{M}$ Tris	262	6.5	P	277
Poly (F-T)		7.9	0.01 $\underline{M}$ Tris	270	4.9	P	277
Poly (F-U)		7.9	0.01 $\underline{M}$ Tris	266	5.2	P	277
Poly (F-fl^{5}U)		7.9	0.01 $\underline{M}$ Tris	274	4.6	P	277
Poly (F-br^{5}U)		7.9	0.01 $\underline{M}$ Tris	280	4.7	P	277
Poly (F-Q)		7.9	0.01 $\underline{M}$ Tris	262	3.8	P	277
Poly (C,br^{5}C)	25	7.0	0.2 $\underline{M}$ cacd.	276	5.09	50%C;257,4.34^k	97
	25	7.0	0.2 $\underline{M}$ cacd.	270	5.44	80%C;258,4.45^k	97
Poly (C,ho^{4}C)		7.7	0.01 $\underline{M}$ Tris			100%C; hper.35%	120
		7.7	0.01 $\underline{M}$ Tris	(270)		80%C;hper.23.8%	120
		7.7	0.01 $\underline{M}$ Tris	(232;272)		60%C;hper.15.3%	120

Polymer	T °C	pH	Medium	λ mμ	$E \times 10^{-3}$ 1/mol·cm	Note	Ref.
Poly (C,ho^{4}C) (Con't)		7.7	0.01 M Tris			0%C;hper.2.0%	120
		7.7	0.01 M Tris	270*		60%C;hper.15.3%[1]	120
		7.7	0.01 M Tris	270*		74%C;hper.24.6%[1]	120
		7.7	0.01 M Tris	270*		80%C;hper.23.8%[1]	120
Poly (C-G)	22	7.8	1 mM phosp.,0.1 mM EDTA	257	6.6	P;hper.(hx-c)18%, 21%(255)	122
Poly (C,G)	25	6.7	0.1 M NaCl,1 mM EDTA, phosp.; I = 0.153 M	259	8.4	C/G=1.3;hpo.22%, 0.526[g]	118
	25	6.7	0.1 M NaCl,1 mM EDTA, phosp.; I = 0.153 M	260	8.4	C/G=1.3;hpo.21%	118
		7.0	0.01 M K phosp.	265	8.2	P; C/G = 16	21
		7.0	0.01 M K phosp.	260	9.4	P; C/G = 2.7	21
Poly d(C,m^{7}G)		2		258		35%C;234 min	86
		7.4		258		35%C;234 min	86
		12		268		35%C;243 min	86
Poly (C-I)		6.5	0.1 M KCl,0.01 M cacd.			hpo.vs.C/I	239
	30.2	7.8	0.09 M NaCl,0.01 M Na phosp.,0.5 mM EDTA	253	4.9	P;hper.(hx-c)40%, 60%(248)	122
	60.5	7.8	0.09 M NaCl,0.01 M Na phosp.,0.5 mM EDTA	248		c.	122
Poly d(C-I)		7.0	0.01 M NaCl,0.1 mM EDTA	251	6.9	P	69
Poly (ho^{4}C,U)		7.8	0.01 M Tris			76%U;hper.7%[1]	120
		7.8	0.01 M Tris			42%U;hper.11%[1]	120
Poly (G,br^{8}G)		7.0	0.0125 M Na cacd.	254		G/brG = 4.56; 0.59,0.16[P]	104
		7.0	0.0125 M Na cacd.	255		G/brG = 3.55; 0.59,0.17[P]	104

Polymer	T °C	pH	Medium	λ mμ	E x 10^{-3} 1/mol·cm	Note	Ref.
Poly (G,br^{8}G) (Con't)		7.9	Tris	260	11.2	23%G	179
Poly (G,sba^{8}G)		7.0	0.01 M NaCl	255;360	10;(6.5)	70%G	199
Poly (G,I)	25	6.7	0.1 M NaCl,1 mM EDTA, phosp.; I = 0.153 M	249.5	8.25	G/I = 1; hpo.59%,0.459^8	118
	25	6.7	0.1 M NaCl,1 mM EDTA, phosp.; I = 0.153 M	260*	6.5	G/I = 1;hpo.48%	118
Poly (G,U)	25	6.7	0.1 M NaCl,1 mM EDTA, phosp.; I = 0.153 M	254	9.7	G/U = 1.1; hpo.27%,0.479^8	118
	25	6.7	0.1 M NaCl,1 mM EDTA, phosp.; I = 0.153 M	260*	9.35	G/U = 1.1, hpo.23%	118
		7.0	0.01 M K phosp.	252	8.8	P; U/G=0.53	21
		7.0	0.01 M K phosp.	257	9.5	P; U/G = 2	21
		7.0	0.01 M K phosp.	257	12.7	P; U/G = 2.7	21
		7.0	0.01 M K phosp.	257	10.7	P; U/G = 6.6	21
		7.0	0.01 M K phosp.	257	10.6	P; U/G = 7.8	21
Poly (I-U)		6.5	0.01 M KCl,0.01 M cacd.			hpo. vs. I/U	239
Poly d(T$_{3-9}$A$_{111}$)					8.84	P	13
(dT)$_6$-d(ib^{6}A)$_{\overline{60}}$		7.0		209;272	14.90;10.80	P;E$_{min}$(235)=3.30	79
(dT)$_6$-d(ac^{6}A)$_{\overline{60}}$		7.0		208;271	15.50;11.20	E$_{min}$(234)=3.50	79
(dT)$_6$-d(ac^{4}C)$_{\overline{60}}$		5.0,7.0,8.9		246;296	12.16;6.24	P;E$_{min}$(273)=4.23, (227)=5.92	79
Poly (U,hm^{5}U)		7	SSC	261		U/hmU=2;e.h.; hper.8.0%(37°)	218
		7	SSC	261		U/hmU=4;e.h.; hper.8.0%(37°)	218
		7	SSC	261		U/hmU=9;e.h.; hper.8.0%(37°)	218

Polymer	$T\,^\circ C$	pH	Medium	λ mμ	$E \times 10^{-3}$ 1/mol·cm	Note	Ref.

2.3.3. Sugar-Phosphate Backbone Analogues

Polymer	$T\,^\circ C$	pH	Medium	λ mμ	$E \times 10^{-3}$ 1/mol·cm	Note	Ref.
Poly ($_S$A$_S$U) 6.5			H$_2$O	259		232min	165
Poly ($_S$C-I) 6.5			H$_2$O	262		252sh; 228min	165
Poly ($_S$C$_S$I) 6.5			H$_2$O	252		228min	165

2.3.3.1. Polyvinylanalogues

Polymer	$T\,^\circ C$	pH	Medium	λ mμ	$E \times 10^{-3}$ 1/mol·cm	Note	Ref.
Poly v^{9}Ade			0.05 M HCl	254	9.2		191
		7	0.05 M NaCl	254	9.0		191
	25	7		254	9.58	hpo.42%	123
Poly v^{9}Acrid			0.02 M HCl	(245;361)	(7.1;0.53)		94
		9	NaOH; ·60% diox.	(246;362)	(7.3;0.6)	hpo.20%(362)	94
Poly v^{1}Cyt			0.05 M HCl	278	7.0	o	190
Poly v^{9}Pur			0.05 M HCl	254	9.2		191
		7	0.05 M NaCl	254	9.0		191
Poly v^{1}Ura		7	Tris	263	~6.0	H = 0.27	189
	25		H$_2$O	265			123
			0.1 M NaOH	264	~4.0	H = 0.24	189
Poly v^1(m^{3}Ura)			1.3 M DMSO	264			191

Complex	$T_{\circ C}$	pH	Medium	λ $m\mu$	$E \times 10^{-3}$ $1/mol \cdot cm$	Note	Ref.

2.3.4. Complexes

Poly A Complexes

Complex	$T_{\circ C}$	pH	Medium	λ $m\mu$	$E \times 10^{-3}$ $1/mol \cdot cm$	Note	Ref.
Poly dT	25	7.8	0.1 M Na phosp.	257	6.9	P; d.s.hx.	210
	25	7.8	0.1 M phosp.	261	11.1	P; c.	210
Poly U		5.5	0.05 M KF,1 mM EDTA	257	6.3	259,12.5[f]	70
			0.05 M KF,1 mM EDTA;e.g.	264	9.15	262,12.72[f]	70
		6.8-7.0	0.01 M cacd.	257.5*	7.19		129
		6.8-7.0	0.01 M cacd.	VAR	VAR		129
	25	7.8	0.1 M Na phosp.	257	7.0	P; d.s.hx.	210
	25	7.8	0.1 M Na phosp.	258.5	11.4	P; c.	210
2 Poly U		6.8-7.0	0.01 M cacd.	257.5*	6.40		129
		6.8-7.0	0.01 M cacd.	VAR	VAR		129
2 Poly e^{5}U	14	N	0.02 M Na$^+$	(263)	(6.4)		247
	70	N	0.02 M Na$^+$	(263)	(9.7)		247
Poly dU	18	7.8	0.038 M Na$^+$,phosp.	260*	7.1	hpo.36.5%	286

Poly n^{2}m^{6}A Complexes

Complex	$T_{\circ C}$	pH	Medium	λ $m\mu$	$E \times 10^{-3}$ $1/mol \cdot cm$	Note	Ref.
Poly U		7.9	0.05 M Na phosp.	263 ~280sh	6.86 ~4.68	$E_{min}(244)=4.4$	102

Poly dA Complexes

Complex	$T_{\circ C}$	pH	Medium	λ $m\mu$	$E \times 10^{-3}$ $1/mol \cdot cm$	Note	Ref.
Poly dT	25	7.8	0.1 M Na phosp.	259	6.0	P; d.s.hx.	210
	25	7.8	0.1 M Na phosp.	261	10.1	P; c.	210
		8.0	1 mM Tris	260*	6.97	P;0.83,0.58[h]	14
Poly U	25	7.8	0.1 M Na phosp.	257	6.5	P; d.s.hx.	210
	25	7.8	0.1 M Na phosp.	258.5	10.4	P; c.	210

Complex	T °C	pH	Medium	λ mμ	$E \times 10^{-3}$ 1/mol·cm	Note	Ref.
Poly dA Complexes (Con't)							
Poly dU	18	7.8	0.038 $\underline{M}$ Na^+, phosp.	260*	6.9	hpo.39%	286
2 Poly dU	18	7.8	0.038 $\underline{M}$ Na^+,phosp.	260*	6.4	hpo.41%	286
	18	7.8	0.038 $\underline{M}$ Na^+,phosp.	max.	6.6		286
Poly d(A-C) Complexes							
Poly d(T-G)		6.8	0.02 $\underline{M}$ NaCl	257		231min	283
		12.3		262		236min	283
Poly C Complexes							
Poly G		5.5	0.05 $\underline{M}$ KF,1 m$\underline{M}$ EDTA	259	6.2	253,8.05[f]	70
			0.05 $\underline{M}$ KF,1 m$\underline{M}$ EDTA;e.g.	256	6.35	254,7.25[f]	70
		6.8	0.1 $\underline{M}$ phosp.	(259)	(7.6)	a.h.	81
		7.0	0.056 $\underline{M}$ NaCl,1 m$\underline{M}$ Na citr.	256.5	6.29		148
		7.0	0.1 $\underline{M}$ NaCl,0.01 $\underline{M}$ Na cacd.	263	7.7	254,9.9[f]	70
			0.1 $\underline{M}$ NaCl,0.01 $\underline{M}$ Na cacd.;e.g.	261	6.8	255,9.6[f]	70
			0.5 $\underline{M}$ KOH	(265)	(8.3)		81
				261	7.5	P; 51%G	37
Poly dG				262	6.9	P; 47%G	37
Poly br5C Complexes							
Poly I	25	7.0	0.1 $\underline{M}$ Na^+, 5 m$\underline{M}$ Na cacd.	283	2.97	271,2.86[k]	97
				245	4.92	232,4.48[k]	97
Poly dC Complexes							
Poly G				261	7.4	P; 50%G	37
Poly dG		N		253	7.4	d.s.hx.; hper.(hx-c)21±2%, 61±2%(276)	110
		N		253	9.1	P; c.	110
		8.0		260*	7.95	P;0.99,0.58[h]	14
				253	7.4	P; 55%G	37

Complex	T °C	pH	Medium	λ mμ	$E \times 10^{-3}$ 1/mol·cm	Note	Ref.
Poly dC Complexes (Con't)							
Poly dI	N			245	5.3	P;d.s.hx.; hper.(hx-c) 53±2%	110
	N			246	8.1	P; c.	110
		8.0	1 m$\underline{M}$ Tris	260*	6.34	P;1.20,0.62[h]	14
Poly d(br^{5}C) Complexes							
Poly dI	N			242	6.65	P; d.s.hx.; hper.(hx-c) 45±6%(244)	110
	N			244	7.8	P; c.	110

FOOTNOTES FOR SECTION 2.3

[a] E is based on E_{259} = 10 x 10^3 of poly A (pH 7, 0.1 $\underline{M}$ KCl, 25°).

[b] Aqueous buffer pH; nominal pH in dioxane mixture = 5.08.

[c] Solution slightly opalescent.

[d] E is based on E_{max} x 10^{-3} values at neutral pH of poly A (10.1) and poly dA (9.9).

[e] Based on E_{max} x 10^{-3} of poly A = 8.2 (pH 5) and 10.1 (pH 7) obtained upon hydrolysis by snake venom phosphodiesterase (24 hr, 37°, pH 8.0) using E_{259} (AMP) = 15.4 x 10^3 (pH 7).

[f] Either λ_{max} and/or E_{max} x 10^{-3} of the constituting nucleoside-5'-monophosphate.

[g] A_{280}/A_{260}.

[h] A_{250}/A_{260} and A_{280}/A_{260}.

[i] A_{270}/A_{260} and A_{280}/A_{260}.

[j] n^2m^6Ado (0.05 $\underline{M}$ Na phosp., pH 7.9): E_{max} (280) = 14,000; E_{sh} (263) = 11,000; E_{min} (242) = 5,710.

[k] λ_{min}, E_{min} x 10^{-3}.

[l] Hydrolysis by pancreatic RNase; E x 10^{-3} (0.1 $\underline{N}$ HCl) = 8.9 (CMP) and 12.2 (hn^{5}CMP).

[m] dCTP: E_{280} = 13.0 x 10^3 (pH 2.0), E_{272} = 9.14 x 10^3 (pH 10.2); d(br^{5}CTP): E_{299} = 9.6 x 10^3 (pH 1.7), E_{287} = 6.85 x 10^3 (pH 7.5).

[n] See the reference for λ_{max}, E_{max}, λ_{min}, and E_{min} of 7-methyl-guanine and its nucleoside and nucleoside mono-, di-, and triphosphates.

o Polymer contained 10% of uracil residues due to hydrolysis of
  cytosine. The E_{max} with different samples varied in the range
  of 6,500-7,200.

p A_{280}/A_{260}, A_{290}/A_{260}.

r The value of 23% of G is based on A_{253}/A_{270} ratio; the chemical
  analysis indicated 40% G.

s E_{280} of cytosine-2'(3')-monophosphate = 13.0×10^3 (pH 2.0).

t 2,4-dithiouridine (53a): E_{max} (280) = 21.2×10^3; E_{max} (340) =
  10.6×10^3 at pH 5.70.

3. MELTING TEMPERATURES

INTRODUCTION

Melting temperature, T_m, is defined as a temperature at which the temperature-induced change of an experimental quantity is 50% complete; it is the equilibrium temperature of transition reactions which are alternatively referred to as, e.g., ordered state $\rightleftharpoons$ disordered state; helix $\rightleftharpoons$ coil; structured form $\rightleftharpoons$ melted form; stacked unit $\rightleftharpoons$ unstacked unit; or complex $\rightleftharpoons$ components. For analysis of melting curves, see ref. 208. Percent thermal hyperchromicity (hper.) accompanying the structured form $\rightleftharpoons$ melted form transition is defined as hper. $= [A_m(\lambda)/A_s(\lambda) - 1] \cdot 100$, where $A_m(\lambda)$ and $A_s(\lambda)$ are absorbances at the wavelength λ of the melted and structured forms of the polymer, respectively. The A_m and A_s values are usually obtained from melting profiles. The melting profiles generally show a sigmoid curve which tends to level off at sufficiently high and low temperatures, where the absorbances are considered to be those of the melted and structured forms, respectively. The temperature interval in which the hyperchromicity was determined may be given in parentheses, e.g., hper.17% (20-90°), particularly in cases where the temperatures of the melted and structured forms cannot be unambiguously determined due to poor definition of the melting profile. Thermal transition, t, usually refers to the steep part of the melting profile; individual steps in multiple transitions are identified by a subscript (e.g., t_1, t_2, ...), where the lower number refers to the lower-temperature transition. For transitions which are incomplete below certain temperatures, such temperatures may be given (e.g., t.incompl. 90°). Broad thermal transition (b.t.) is usually indicative of a noncooperative transition.

Dissociation (melting) or association of oligo- and polynucleotide structures in a thermal transition step, which is characterized by the T_m value, may be shown in an abbreviated form as follows: $3 \rightarrow 1$ represents a dissociation of a triple-stranded complex into single-stranded components (e.g., poly A·2 poly U $\rightarrow$ poly A + 2 poly U); $2 \rightarrow 3$ represents formation of a triple-stranded complex from a double-stranded one (e.g., poly A·poly U + poly U $\rightarrow$ poly A·2 poly U); etc. Transitions of other types are formulated analogously.

The methods of determining T_m values are abbreviated as follows:

CAL   Calorimetry

CD   Circular dichroism; the wavelength ($m\mu$) at which ellipticity was measured may be given following CD (e.g., CD270). CDmax refers to the maximum of the circular dichroic spectrum.

COND   Conductometry.

EB   Electrical birefringence.

IMMUN   Immunochemical measurements; T_m was determined from a plot of complement-fixing reactivity of a poly-nucleotide complex vs. temperature (244); such plots were very similar in shape to the absorbance-temperature profiles.

IR   Infrared spectrophotometry; wave number (cm^{-1}) at which optical density was measured may replace IR (e.g., 1780).

OD   Optical density in the ultraviolet region; wavelength ($m\mu$) at which optical density was measured may replace OD; max refers to the wavelength of the absorption maximum.

ORD   Optical rotatory dispersion; the wavelength ($m\mu$) at which rotation was measured may be given following ORD (e.g., ORD289).

3. MELTING TEMPERATURES

3.1. OLIGONUCLEOTIDES

Oligomer	Method	pH	Medium	T_m ^oC	Notes	Ref.

3.1.1. Homooligomers

Oligomer	Method	pH	Medium	T_m ^oC	Notes	Ref.
A-A	ORD	7	25.2% LiCl	5		263
	259	7	0.1 $\underline{M}$ LiCl,0.01 $\underline{M}$ cacd.	(34)	nco.	135
	CDmax	7.4	0.1 $\underline{M}$ NaCl,0.1 $\underline{M}$ Tris	(24)		275
	CDmax	7.5	4.7 $\underline{M}$ KF,0.01 $\underline{M}$ Tris	27		19
	CD	7.5	4.7 $\underline{M}$ KF,0.01 $\underline{M}$ Tris	21		11
$(Ap)_6A$	259	7	0.1 $\underline{M}$ LiCl,0.01 $\underline{M}$ cacd.	(44)	nco.	135
$(Ap)_2A$	CD	7.4	0.1 $\underline{M}$ NaCl,0.01 $\underline{M}$ Tris	(14)	nco.	20
$(Ap)_4A$	CD	7.4	0.1 $\underline{M}$ NaCl,0.01 $\underline{M}$ Tris	(18)	nco.	20
$(Ap)_6A$	CD	7.4	0.1 $\underline{M}$ NaCl,0.01 $\underline{M}$ Tris	(26)	nco.	20
$(Ap)_{11}A$	CD	7.4	0.1 $\underline{M}$ NaCl,0.01 $\underline{M}$ Tris	(30)	nco.	20
Poly A	CD	7.4	0.1 $\underline{M}$ NaCl,0.01 $\underline{M}$ Tris	(41)	nco.	20
$(Ap)_nA$	260	4.0	0.1 $\underline{M}$ NaCl,0.05 $\underline{M}$ acet.	VAR	n = 5-59; co.	20
$(Ap)_nA$	260	4.5	0.1 $\underline{M}$ NaCl,0.05 $\underline{M}$ acet.	VAR	n = 6-199; co.	20
$(Ap)_nA$	260	4.9	0.1 $\underline{M}$ NaCl,0.05 $\underline{M}$ acet.	VAR	n = 7-199; co.	20
A-Ap	CDmax	7.5	4.7 $\underline{M}$ KF,0.01 $\underline{M}$ Tris	4		19
pA-A	ORD259	6.8	0.15 $\underline{M}$ KCl,0.04 $\underline{M}$ NaH$_2$PO$_4$, 0.06 $\underline{M}$ Na$_2$HPO$_4$	(30)		201
	CDmax	7.5	4.7 $\underline{M}$ KF,0.01 $\underline{M}$ Tris	27		19
$(pA)_6$	ORD257	6.8	0.15 $\underline{M}$ KCl,0.04 $\underline{M}$ NaH$_2$PO$_4$, 0.06 $\underline{M}$ Na$_2$HPO$_4$	(34)		201

Oligomer	Method	pH	Medium	T_m °C	Notes	Ref.
$(pA)_7$	260	4.2	0.1 $\underline{M}$ NaClO$_4$,citr.,phosp.; I = $\overline{0}$.12 $\underline{M}$	29±0.5	0.05 $\underline{mM}$ A	53
$(pA)_8$	265	4.2	0.1 $\underline{M}$ NaClO$_4$,citr.,phosp.; I = $\overline{0}$.12 $\underline{M}$	(40)	0.05 $\underline{mM}$ A	53
$(pA)_9$	265	4.2	0.1 $\underline{M}$ NaClO$_4$,citr.,phosp.; I = $\overline{0}$.12 $\underline{M}$	(47)	0.05 $\underline{mM}$ A	53
$(pA)_{10}$	260	4.2	0.1 $\underline{M}$ NaClO$_4$,citr.,phosp.; I = $\overline{0}$.12 $\underline{M}$	53.3	0.1 $\underline{mM}$ A	53
$(pA)_n$	OD	4.2	0.1 $\underline{M}$ NaClO$_4$,citr.,phosp.; I = $\overline{0}$.12 $\underline{M}$	VAR	n = 6-10; VAR hper.	53
$(pA)_5$	OD	VAR	0.1 $\underline{M}$ NaClO$_4$,citr.,phosp.; I = $\overline{0}$.12 $\underline{M}$	VAR		53
$(pA)_6$	OD	VAR	0.1 $\underline{M}$ NaClO$_4$,citr.,phosp.; I = $\overline{0}$.12 $\underline{M}$	VAR		53
A_5		4.00	SSC	11.8		5
A_6		4.00	SSC	23.4		5
	260	4.0	0.1 $\underline{M}$ NaCl,0.05 $\underline{M}$ Na acet.	~8		177
A_7	260	4.0	0.1 $\underline{M}$ NaCl,0.05 $\underline{M}$ Na acet.	~22		177
	260	4.5	0.1 $\underline{M}$ NaCl,0.05 $\underline{M}$ Na acet.	~3		177
A_8		4.00	SSC	41.4		5
	260	4.0	0.1 $\underline{M}$ NaCl,0.05 $\underline{M}$ Na acet.	38		177
	260	4.5	0.1 $\underline{M}$ NaCl,0.05 $\underline{M}$ Na acet.	~19		177
	260	4.86	0.1 $\underline{M}$ NaCl,0.05 $\underline{M}$ Na acet.	~3		177
A_9		4.00	SSC	46.9		5
A_{10}		4.00	SSC	53.2		5

Oligomer	Method	pH	Medium	T_m °C	Notes	Ref.
A_{11}		4.00	SSC	58.5		5
A_{12}	260	4.0	0.1 M NaCl,0.05 M Na acet.	71.1		177
	260	4.5	0.1 M NaCl,0.05 M Na acet.	52		177
	260	4.86	0.1 M NaCl,0.05 M Na acet.	36		177
A_{30}	260	4.0	0.1 M NaCl,0.05 M Na acet.	88.1		177
	260	4.5	0.1 M NaCl,0.05 M Na acet.	66.4		177
	260	4.86	0.1 M NaCl,0.05 M Na acet.	52.3		177
A_{45}	260	4.0	0.1 M NaCl,0.05 M Na acet.	92.1		177
	260	4.5	0.1 M NaCl,0.05 M Na acet.	71		177
	260	4.86	0.1 M NaCl,0.05 M Na acet.	57.2		177
A_{60}	260	4.0	0.1 M NaCl,0.05 M Na acet.	97.4		177
	260	4.5	0.1 M NaCl,0.05 M Na acet.	73		177
	260	4.86	0.1 M NaCl,0.05 M Na acet.	58.6		177
A_{200}	260	4.0	0.1 M NaCl,0.05 M Na acet.	>100		177
	260	4.5	0.1 M NaCl,0.05 M Na acet.	77.2		177
	260	4.86	0.1 M NaCl,0.05 M Na acet.	64.3		177
Am-Am	CD	7.5	4.7 M KF,0.01 M Tris	15		11
C-C	CDmax	7.5	4.7 M KF,0.01 M Tris	24		17,19
	CDmax	7.5-8.0	4.7 M KF,0.01 M Tris	(20)	nco.	18
$(Cp)_2C$	CDmax	7.5-8.0	4.7 M KF,0.01 M Tris	(32)	nco.	18
$(Cp)_3C$	CDmax	7.5-8.0	4.7 M KF,0.01 M Tris	(34)	nco.	18
$(Cp)_4C$	CDmax	7.5-8.0	4.7 M KF,0.01 M Tris	(32)	nco.	18
$(Cp)_9C$	CDmax	7.5-8.0	4.7 M KF,0.01 M Tris	(58)	nco.	18
	CDmax	7.5	4.7 M KF,0.01 M Tris	59		17

Oligomer	Method	pH	Medium	T_m °C	Notes	Ref.
$(Cp)_nC$	280	4.0	0.1 $\underline{M}$ NaCl, 0.05 $\underline{M}$ Na acet.	VAR	n = 7-13; co.	18
C_6	280	4.0	0.1 $\underline{M}$ NaCl, 0.05 $\underline{M}$ Na acet.	>0		177
C_7	280	4.0	0.1 $\underline{M}$ NaCl, 0.05 $\underline{M}$ Na acet.	20		177
C_8	280	4.0	0.1 $\underline{M}$ NaCl, 0.05 $\underline{M}$ Na acet.	25.6		177
C_9	280	4.0	0.1 $\underline{M}$ NaCl, 0.05 $\underline{M}$ Na acet.	34.7		177
C_{10}	280	4.0	0.1 $\underline{M}$ NaCl, 0.05 $\underline{M}$ Na acet.	41.5		177
C_{13}	280	4.0	0.1 $\underline{M}$ NaCl, 0.05 $\underline{M}$ Na acet.	58.0		177
C2'-5'C	CDmax	7.5	0.1 $\underline{M}$ NaCl, 0.05 $\underline{M}$ Na acet.		no t. > -20°	19
d(C-C)	CDmax	7.5	0.1 $\underline{M}$ NaCl, 0.05 $\underline{M}$ Na acet.		no t. > -20°	19
a(C-C)	CDmax	7.5	0.1 $\underline{M}$ NaCl, 0.05 $\underline{M}$ Na acet.		no t. > -20°	19
G-G	260	6.2	0.2 $\underline{M}$ NaCl, 2 m$\underline{M}$ Na cacd.	17.4	0.46 m$\underline{M}$ G[a,b]	146
	260	7.2	0.2 $\underline{M}$ NaCl, 2 m$\underline{M}$ Na cacd.	18.5	0.378 m$\underline{M}$ G[a]	146
	260	8.8	0.2 $\underline{M}$ NaCl, 2 m$\underline{M}$ Na cacd.	16.4	0.378 m$\underline{M}$ G[a]	146
G-G-G	260	6.2	0.2 $\underline{M}$ NaCl, 2 m$\underline{M}$ Na cacd.	23.6	0.42 m$\underline{M}$ G	146
	260	7.4	0.2 $\underline{M}$ NaCl, 2 m$\underline{M}$ Na cacd.	21.8	0.09 m$\underline{M}$ G	146
	260	7.4	0.2 $\underline{M}$ NaCl, 2 m$\underline{M}$ Na cacd.	24.3	0.42 m$\underline{M}$ G	146
	260	8.8	0.2 $\underline{M}$ NaCl, 2 m$\underline{M}$ Na cacd.	23.5	0.042 m$\underline{M}$ G	146
	260	8.8	0.2 $\underline{M}$ NaCl, 2 m$\underline{M}$ Na cacd.	28.4	0.42 m$\underline{M}$ G	146
G2'-5'G	260	6.2	0.2 $\underline{M}$ NaCl, 2 m$\underline{M}$ Na cacd.	<15	0.585 m$\underline{M}$ G	146
G-Gp	260	5.0	0.2 $\underline{M}$ acet.	26.2	4.32 m$\underline{M}$ G	146
	260	6.2	0.2 $\underline{M}$ NaCl, 2 m$\underline{M}$ Na cacd.	22.3	4.32 m$\underline{M}$ G[c]	146
$(Gp)_3$	240	6.2	0.2 $\underline{M}$ NaCl, 2 m$\underline{M}$ cacd.	23.7		145
	260	6.2	0.2 $\underline{M}$ NaCl, 2 m$\underline{M}$ Na cacd.	23.7	0.42 m$\underline{M}$ G	146

Oligomer	Method	pH	Medium	T_m °C	Notes	Ref.
$(Up)_{14}$	260	6.6	0.5 $\underline{M}$ CsCl, 0.05 $\underline{M}$ Na cacd., 0.1 $\underline{mM}$ EDTA	−4.5		49a
	260	6.6	5 $\underline{mM}$ Na cacd., 0.025 $\underline{mM}$ sperm.	2		49a
$(Up)_{17}U$	260	6.6	0.5 $\underline{M}$ CsCl, 0.05 $\underline{M}$ Na cacd., 0.1 $\underline{mM}$ EDTA	−1		49a
	260	6.6	5 $\underline{mM}$ Na cacd., 0.025 $\underline{mM}$ sperm.	6.4		49a
$(Up)_{23}U$	260	6.6	0.5 $\underline{M}$ CsCl, 0.05 $\underline{M}$ Na cacd., 0.1 $\underline{mM}$ EDTA	2.5		49a
	260	6.6	5 $\underline{mM}$ Na cacd., 0.025 $\underline{mM}$ sperm.	11		49a
$(Up)_{30}U$	260	6.6	0.5 $\underline{M}$ CsCl, 0.05 $\underline{M}$ Na cacd., 0.01 $\underline{mM}$ EDTA	5.6		49a
	260	6.6	5 $\underline{mM}$ Na cacd., 0.025 $\underline{mM}$ sperm.	15		49a
Poly U	260	6.6	0.5 $\underline{M}$ CsCl, 0.05 $\underline{M}$ Na cacd., 0.01 $\underline{mM}$ EDTA	10.5		49a
	260	6.6	5 $\underline{mM}$ Na cacd., 0.025 $\underline{mM}$ sperm.	31		49a

Oligomer	Method	pH	Medium	T_m °C	Notes	Ref.

3.1.2. Heterooligomers

Oligomer	Method	pH	Medium	T_m °C	Notes	Ref.
A–A–C	CD	7.4	4.7 $\underline{M}$ KF, 0.01 $\underline{M}$ Tris	(38)		16
	CD	7.4	0.01 $\underline{M}$ Tris	(22)		16
A–A–Cp	CD	7.4	4.7 $\underline{M}$ KF, 0.01 $\underline{M}$ Tris	(48)		16
A–C	CDmax	7.5	4.7 $\underline{M}$ KF,0.01 $\underline{M}$ Tris	25		17,19
A–aC	CDmax	7.5	4.7 $\underline{M}$ KF,0.01 $\underline{M}$ Tris	35		19
d(A–T)$_5$	262	(N)	0.51 $\underline{M}$ Na$^+$	($\sim$16);(65)		217
d(A–T)$_{13}$	262	(N)	0.51 $\underline{M}$ Na$^+$	(34);(58)		217
d(A–T)$_{21}$	262	(N)	0.51 $\underline{M}$ Na$^+$	(43);(65)		217
d(A͡–T)$_{\underline{20}}$	262	(N)	< 0.01 $\underline{M}$ Na$^+$	(57)		217
	262	(N)	0.51 $\underline{M}$ Na$^+$	(75)		217
d(A–T)$_{\underline{31}}$	262	(N)	< 0.01 $\underline{M}$ Na$^+$	(52)		217
	262	(N)	0.51 $\underline{M}$ Na$^+$	(74)		217
A–U	CDmax	7.5	4.7 $\underline{M}$ KF,0.01 $\underline{M}$ Tris	11		17
C–A	CDmax	7.5	4.7 $\underline{M}$ KF,0.01 $\underline{M}$ Tris	15		17,19
aC–A	CDmax	7.5	4.7 $\underline{M}$ KF,0.01 $\underline{M}$ Tris		no t. > -20°	19
C–U	CDmax	7.5	4.7 $\underline{M}$ KF,0.01 $\underline{M}$ Tris	6		17
G–A	CDmax	7.5	4.7 $\underline{M}$ KF,0.01 $\underline{M}$ Tris	9		17
G–A–U	CD	7.4	4.7 $\underline{M}$ KF,0.01 $\underline{M}$ Tris	(19)		16
G–A–Up	CD	7.4	4.7 $\underline{M}$ KF,0.01 $\underline{M}$ Tris	(27)		16

Complex	Method	pH	Medium	T_m °C	Note	Ref.

3.1.3. Oligonucleotide Complexes and Mixtures

Complex	Method	pH	Medium	T_m °C	Note	Ref.
$(Ap)_5 A \cdot 2 (Up)_5 U$	260	7.30	0.5 $\underline{M}$ NaCl, 0.04 $\underline{M}$ phosp.	(11)		29
$d(pA)_6 \cdot d(pT)_{12}$	259	7	1 $\underline{M}$ NaCl, 0.01 $\underline{M}$ Na phosp.	16	d	186
$(Ap)_6 C, G(pU)_6$	CD	7.0-7.5	1 $\underline{M}$ NaCl, 0.04 $\underline{M}$ phosp.	25	10 m$\underline{M}$ N	65
$(Ap)_6 C, 2 G(pU)_6$	CD	7.0-7.5	1 $\underline{M}$ NaCl, 0.04 $\underline{M}$ phosp.	25	10 m$\underline{M}$ N	65
$(Ap)_7 C, G(pU)_7$	CD	7.0-7.5	1 $\underline{M}$ NaCl, 0.04 $\underline{M}$ phosp.	21	1 m$\underline{M}$ N	65
	CD	7.0-7.5	1 $\underline{M}$ NaCl, 0.04 $\underline{M}$ phosp.	22.5	2 m$\underline{M}$ N	65
	CD	7.0-7.5	1 $\underline{M}$ NaCl, 0.04 $\underline{M}$ phosp.	31.5	10 m$\underline{M}$ N	65
$(Ap)_7 C, 2 G(pU)_7$	CD	7.0-7.5	1 $\underline{M}$ NaCl, 0.04 $\underline{M}$ phosp.	30	10 m$\underline{M}$ N	65
$(Ap)_6 C, (Up)_6 G$	CD	7.0-7.5	1 $\underline{M}$ NaCl, 0.04 $\underline{M}$ phosp.	18.5	10 m$\underline{M}$ N	65
$(Ap)_6 C, 2 (Up)_6 G$	CD	7.0-7.5	1 $\underline{M}$ NaCl, 0.04 $\underline{M}$ phosp.	17	10 m$\underline{M}$ N	65
$(Ap)_6 C, G(pU)_6, (Up)_6 G$	CD	7.0-7.5	1 $\underline{M}$ NaCl, 0.04 $\underline{M}$ phosp.	20.5	10 m$\underline{M}$ N	65
$d(A-C)_n \cdot d(G-T)_m$	265	7.0	0.02 $\underline{M}$ Na$^+$, phosp.	VAR	VAR n,m, and concn.[e]	84
	265	7.0	0.07 $\underline{M}$ Na$^+$, phosp.	VAR	VAR n,m, and concn.[e]	84
$C(pA)_6, G(pU)_6$	CD	7.0-7.5	1 $\underline{M}$ NaCl, 0.04 $\underline{M}$ phosp.	13.5	10 m$\underline{M}$ N	65
$C(pA)_6, 2 G(pU)_6$	CD	7.0-7.5	1 $\underline{M}$ NaCl, 0.04 $\underline{M}$ phosp.	13	10 m$\underline{M}$ N	65
$C(pA)_7, G(pU)_7$	CD	7.0-7.5	1 $\underline{M}$ NaCl, 0.04 $\underline{M}$ phosp.	23	10 m$\underline{M}$ N	65
$C(pA)_7, 2 G(pU)_7$	CD	7.0-7.5	1 $\underline{M}$ NaCl, 0.04 $\underline{M}$ phosp.	22	10 m$\underline{M}$ N	65
$C(pA)_6, (Up)_6 G$	CD	7.0-7.5	1 $\underline{M}$ NaCl, 0.04 $\underline{M}$ phosp.	19	10 m$\underline{M}$ N	65
$C(pA)_6, 2 (Up)_6 G$	CD	7.0-7.5	1 $\underline{M}$ NaCl, 0.04 $\underline{M}$ phosp.	18	10 m$\underline{M}$ N	65

Complex	Method	pH	Medium	T_m °C	Note	Ref.
$C(pA)_6, G(pU)_6, (Up)_6 G$	CD	7.0-7.5	1 $\underline{M}$ NaCl, 0.04 $\underline{M}$ phosp.	15.5	10 m$\underline{M}$ N	65
G-C-C.2 G-G-C	ORD275	7.0	0.05 $\underline{M}$ Na phosp., 0.01 $\underline{M}$ MgCl$_2$	(23)		263
	ORD292	7.0	0.05 $\underline{M}$ Na phosp., 0.01 $\underline{M}$ MgCl$_2$	(25)		263

FOOTNOTES FOR SECTION 3.1

[a] No hyperchromic effect in 0.042 mM solution.

[b] See refs. 66 and 214a for T_m values of GMP gels.

[c] No hyperchromic effect in 0.70 mM solution.

[d] See the reference for hypochromicities (260 mμ, $0°$, 1 M NaCl + 0.01 M Na phosphate, pH 6.3) of mixtures of $d(pA)_m$ (m = 3-9) and $d(pT)_n$ (n = 4-10).

[e] Experimental data give the relationship $T_m = 79.9 - 498 \cdot n^{-1}$ (for 0.02 M Na$^+$) and $T_m = 87.5 - 442 \cdot n^{-1}$ (for 0.07 M Na$^+$), where n is the average length in nucleotide units of the shorter member of the mixture oligo d(A-C) + oligo d(G-T). T_m values are tabulated for various concentrations and ratios of oligo d(A-C) and oligo d(G-T) mixtures.

3.2. COMPLEXES OF POLYNUCLEOTIDES WITH THEIR COMPONENTS

Complex	Method	pH	Medium	T_m °C	Note	Ref.
Poly A Complexes						
$d(pT)_6$		6.3	1 $\underline{M}$ NaCl,0.01 $\underline{M}$ Na phosp.	(10)	a	186
$d(pT)_n$	OD	7.0	0.04 $\underline{M}$ K phosp.	VAR	VAR n	34,35
	OD	7.0	0.04 $\underline{M}$ K phosp.,8 m$\underline{M}$ MgCl$_2$	VAR	VAR n	34,35
	OD	7.0	SSC	VAR	VAR n	34,35
U_6	260	7.0	0.1 $\underline{M}$ NaCl,0.05 $\underline{M}$ Na cacd., 0.01 $\underline{M}$ Mg^{++}	<0		177
U_7	260	7.0	0.1 $\underline{M}$ NaCl,0.05 $\underline{M}$ Na cacd., 0.01 $\underline{M}$ Mg^{++}	<9.6		177
U_8	260	7.0	0.1 $\underline{M}$ NaCl,0.05 $\underline{M}$ Na cacd., 0.01 $\underline{M}$ Mg^{++}	20		177
U_{10}	260	7.0	0.1 $\underline{M}$ NaCl,0.05 $\underline{M}$ Na cacd., 0.01 $\underline{M}$ Mg^{++}	37.1		177
U_{12}	260	7.0	0.1 $\underline{M}$ NaCl,0.05 $\underline{M}$ Na cacd., 0.01 $\underline{M}$ Mg^{++}	51.4		177
U_{13}	260	7.0	0.1 $\underline{M}$ NaCl,0.05 $\underline{M}$ Na cacd., 0.01 $\underline{M}$ Mg^{++}	58.3		177
U_{15}	260	7.0	0.1 $\underline{M}$ NaCl,0.05 $\underline{M}$ Na cacd., 0.01 $\underline{M}$ Mg^{++}	68.1		177
U_{16}	260	7.0	0.1 $\underline{M}$ NaCl,0.05 $\underline{M}$ Na cacd., 0.01 $\underline{M}$ Mg^{++}	69.5		177
$(br^5U)_n$		6.7	SSC	24	A/brU = 1;n 7	254
Poly dA Complexes						
$d(pT)_7$	260	7.0	0.04 $\underline{M}$ K phosp.,8 m$\underline{M}$ MgCl$_2$	14.5	0.12[b]	34
	260	N	SSC	9.0	0.16[b]	34

Complex	Method	pH	Medium	T_m $^\circ C$	Note	Ref.
Poly dA Complexes (Con't)						
$d(pT)_8$	260	7.0	0.04 $\underline{M}$ K phosp.,8 m$\underline{M}$ MgCl$_2$	21.0	0.12[b]	34
	OD	N	SSC	VAR	Var concn. of $d(pT)_8$	34
$d(pT)_9$	260	7.0	0.04 $\underline{M}$ K phosp.	14.0	0.14[b]	34
	260	N	SSC	20.8	0.14[b]	34
$d(pT)_{10}$	260	7.0	0.04 $\underline{M}$ K phosp.	17.0	0.15[b]	34
	260	7.0	0.04 $\underline{M}$ K phosp.,8 m$\underline{M}$ MgCl$_2$	29.2	0.10[b]	34
	260	N	SSC	25.8	0.14[b]	34
$d(pT)_{11}$	260	7.0	0.04 $\underline{M}$ K phosp.	20.5	0.14[b]	34
	260	7.0	0.04 $\underline{M}$ K phosp.,8 m$\underline{M}$ MgCl$_2$	33.6	0.12[b]	34
	260	N	SSC	30.0	0.10[b]	34
$d(pT)_{12}$	260	7.0	0.04 $\underline{M}$ K phosp.	23.0	0.14[b]	34
	260	7.0	0.04 $\underline{M}$ K phosp.,8 m$\underline{M}$ MgCl$_2$	35.7	0.10[b]	34
$d(pT)_{13}$	260	7.0	0.04 $\underline{M}$ K phosp.	25.5	0.14[b]	34
	260	7.0	0.04 $\underline{M}$ K phosp.,8 m$\underline{M}$ MgCl$_2$	38.5	0.10[b]	34
	260	N	SSC	36.0	0.11[b]	34
$d(pT)_{15}$	260	7.0	0.04 $\underline{M}$ K phosp.	28.5	0.14[b]	34
	260	7.0	0.04 $\underline{M}$ K phosp.,8 m$\underline{M}$ MgCl$_2$	42.2	0.10[b]	34
	260	N	SSC	40.5	0.09[b]	34
$d(pT)_{17}$	260	7.0	0.04 $\underline{M}$ K phosp.	34.3	0.09[b]	34
	260	7.0	0.04 $\underline{M}$ K phosp.,8 m$\underline{M}$ MgCl$_2$	46.8	0.12[b]	34
$d(pT)_{18}$	260	7.0	0.04 $\underline{M}$ K phosp.	34.8	0.10[b]	34
	260	7.0	0.04 $\underline{M}$ K phosp.,8 m$\underline{M}$ MgCl$_2$	47.5	0.12[b]	34
$d(pT)_{19}$	260	7.0	0.04 $\underline{M}$ K phosp.	38.0	0.10[b]	34

Complex	Method	pH	Medium	T_m °C	Note	Ref.

Poly dA Complexes (Con't)

d(pT)$_{19}$ (Con't)

Complex	Method	pH	Medium	T_m °C	Note	Ref.
	260	7.0	0.04 M K phosp.,8 mM MgCl$_2$	48.8	0.12[b]	34
	260	N	SSC	44.5	0.09[b]	34

Poly isoA Complexes

Complex	Method	pH	Medium	T_m °C	Ref.
I$_4$	280	7.0	0.02 M NaCl,0.01 M cacd.	< 0	177
	280	7.0	0.02 M NaCl,0.01 M cacd., 0.01 M Mg^{++}	12	177
I$_5$	280	7.0	0.02 M NaCl,0.01 M cacd.	9.8	177
	280	7.0	0.02 M NaCl,0.01 M cacd., 0.01 M Mg^{++}	25.9	177
I$_6$	280	7.0	0.02 M NaCl,0.01 M cacd.	14.7	177
	280	7.0	0.02 M NaCl,0.01 M cacd., 0.01 M Mg^{++}	41.3	177
I$_7$	280	7.0	0.02 M NaCl,0.01 M cacd.	~28.5	177
	280	7.0	0.02 M NaCl,0.01 M cacd., 0.01 M Mg^{++}	>73.7	177
I$_9$	280	7.0	0.02 M NaCl,0.01 M cacd.	~59.0	177
	280	7.0	0.02 M NaCl,0.01 cacd., 0.01 M Mg^{++}	>85.8	177
I$_{11}$	280	7.0	0.02 M NaCl,0.01 M cacd.	~72.0	177
	280	7.0	0.02 M NaCl,0.01 M cacd., 0.01 M Mg^{++}	>89	177
U$_3$	270	7.0	0.1 M NaCl,0.05 M cacd., 0.01 M Mg^{++}	< 0	177
U$_4$	270	7	0.02 M NaCl,0.01 M cacd.	< 0	177
	270	7.0	0.1 M NaCl,0.05 M cacd., 0.01 M Mg^{++}	13.9	177
U$_5$	270	7	0.02 M NaCl,0.01 M cacd.	< 8.5	177

Complex	Method	pH	Medium	T_m °C	Note	Ref.

Poly isoA Complexes (Con't)

U_5 (Con't)

	270	7.0	0.1 $\underline{M}$ NaCl, 0.05 $\underline{M}$ cacd., 0.01 $\underline{M}$ Mg^{++}	33.9		177
U_6	270	7	0.02 $\underline{M}$ NaCl, 0.01 $\underline{M}$ cacd.	< 16.4		177
	270	7.0	0.1 $\underline{M}$ NaCl, 0.05 $\underline{M}$ cacd., 0.01 $\underline{M}$ Mg^{++}	41.3		177
U_7	270	7	0.02 $\underline{M}$ NaCl, 0.01 $\underline{M}$ cacd.	24.8		177
	270	7.0	0.1 $\underline{M}$ NaCl, 0.05 $\underline{M}$ cacd., 0.01 $\underline{M}$ Mg^{++}	47.6		177
U_8	270	7	0.02 $\underline{M}$ NaCl, 0.01 $\underline{M}$ cacd.	35.5		177
	270	7.0	0.1 $\underline{M}$ NaCl, 0.05 $\underline{M}$ cacd., 0.01 $\underline{M}$ Mg^{++}	70.2		177
U_{10}	270	7	0.02 $\underline{M}$ NaCl, 0.01 $\underline{M}$ cacd.	53.3		177
	270	7.0	0.1 $\underline{M}$ NaCl, 0.05 $\underline{M}$ cacd., 0.01 $\underline{M}$ Mg^{++}	82.0		177
U_{12}	270	7	0.02 $\underline{M}$ NaCl, 0.01 $\underline{M}$ cacd.	69.2		177
	270	7.0	0.1 $\underline{M}$ NaCl, 0.05 $\underline{M}$ cacd., 0.01 $\underline{M}$ Mg^{++}	>93.0		177
U_{13}	270	7	0.02 $\underline{M}$ NaCl, 0.01 $\underline{M}$ cacd.	74.7		177
	270	7.0	0.1 $\underline{M}$ NaCl, 0.05 $\underline{M}$ cacd., 0.01 $\underline{M}$ Mg^{++}	>97.0		177
U_{15}	270	7	0.02 $\underline{M}$ NaCl, 0.01 $\underline{M}$ cacd.	80.4		177
	270	7.0	0.1 $\underline{M}$ NaCl, 0.05 $\underline{M}$ cacd., 0.01 $\underline{M}$ Mg^{++}	>100		177
U_{16}	270	7	0.02 $\underline{M}$ NaCl, 0.01 $\underline{M}$ cacd.	85.3		177
	270	7.0	0.1 $\underline{M}$ NaCl, 0.05 $\underline{M}$ cacd., 0.01 $\underline{M}$ Mg^{++}	>100		177

Poly C Complexes

| Guo | ORD292 | 4.5 | 0.01 $\underline{M}$ Na$^+$ | (12) | | 216 |
| | ORD275 | 6.4 | 0.01 $\underline{M}$ Na$^+$ | (34) | | 216 |

Complex	Method	pH	Medium	T °C	Note	Ref.

Poly C Complexes (Con't)

Guo (Con't)

Complex	Method	pH	Medium	T °C	Note	Ref.
	OD	VAR	0.01 and 0.25 $\underline{M}$ Na$^+$	VAR		216
Guo-3'-P	OD	VAR	0.25 $\underline{M}$ Na$^+$	VAR		216
Guo-2'/3'-P	OD	VAR	0.1,0.25, and 1.0 $\underline{M}$ Na$^+$	VAR		216
dGuo-5'-P	OD	4.50	0.01 $\underline{M}$ NaCl	30	C/G = 2	44
	1525,1580	7.0	0.2 $\underline{M}$ Na cacd.; D_2O	(23)	C/G = 2;t.s.hx.	95
G-G	1655	6.7	0.06 $\underline{M}$ Na cacd.; D_2O	(34)	C/G = 2;t.s.hx.	95
$(Gp)_3$	1580,1680	8.3	0.13 $\underline{M}$ Na cacd.; D_2O	(36)	C/G = 2	95
$2(Gp)_3$	240	6.2	0.2 $\underline{M}$ NaCl, 2 m$\underline{M}$ cacd.	24.7;(46)	3→2;2→1^c	145
I_6	245	7.0	0.1 $\underline{M}$ NaCl,0.05 $\underline{M}$ cacd.	<0		177
	245	7.0	0.1 $\underline{M}$ NaCl,0.05 $\underline{M}$ cacd., 0.01 $\underline{M}$ Mg^{++}	5		177
I_7	245	7.0	0.1 $\underline{M}$ NaCl,0.05 $\underline{M}$ cacd.	5.5		177
	245	7.0	0.1 $\underline{M}$ NaCl,0.05 $\underline{M}$ cacd., 0.01 $\underline{M}$ Mg^{++}	16.6		177
I_9	245	7.0	0.1 $\underline{M}$ NaCl,0.05 $\underline{M}$ cacd.	30.3		177
	245	7.0	0.1 $\underline{M}$ NaCl,0.05 $\underline{M}$ cacd., 0.01 $\underline{M}$ Mg^{++}	42.1		177
I_{11}	245	7.0	0.1 $\underline{M}$ NaCl,0.05 $\underline{M}$ cacd.	47.8		177
	245	7.0	0.1 $\underline{M}$ NaCl,0.05 $\underline{M}$ cacd., 0.01 $\underline{M}$ Mg^{++}	58.9		177
Pur	ORD589	4.6	Na acet.; I = 0.2 $\underline{M}$	(69.5)		213

Poly br^{5}C Complexes

Complex	Method	pH	Medium	T °C	Note	Ref.
I_4	250	7.0	0.1 $\underline{M}$ NaCl,0.05 $\underline{M}$ cacd.	<0		177

Complex	Method	pH	Medium	T_m °C	Note	Ref.

Poly br^{5}C Complexes (Con't)

I_4 (Con't)

| | 250 | 7.0 | 0.1 M NaCl, 0.05 M cacd., 0.01 M Mg^{++} | ~5 | | 177 |

I_5

| | 250 | 7.0 | 0.1 M NaCl, 0.05 M cacd. | < 10.8 | | 177 |
| | 250 | 7.0 | 0.1 M NaCl, 0.05 M cacd., 0.01 M Mg^{++} | 18.5 | | 177 |

I_6

| | 250 | 7.0 | 0.1 M NaCl, 0.05 M cacd. | 18.0 | | 177 |
| | 250 | 7.0 | 0.1 M NaCl, 0.05 M cacd., 0.01 M Mg^{++} | 27.8 | | 177 |

I_7

| | 250 | 7.0 | 0.1 M NaCl, 0.05 M cacd. | 31.3 | | 177 |
| | 250 | 7.0 | 0.1 M NaCl, 0.05 M cacd., 0.01 M Mg^{++} | 40.3 | | 177 |

I_9

| | 250 | 7.0 | 0.1 M NaCl, 0.05 M cacd. | 54.5 | | 177 |
| | 250 | 7.0 | 0.1 M NaCl, 0.05 M cacd., 0.01 M Mg^{++} | 63.2 | | 177 |

I_{11}

| | 250 | 7.0 | 0.1 M NaCl, 0.05 M cacd. | 72.6 | | 177 |
| | 250 | 7.0 | 0.1 M NaCl, 0.05 M cacd., 0.01 M Mg^{++} | 79.7 | | 177 |

Poly I Complexes

C_7

| | 245 | 7.0 | 0.1 M NaCl, 0.05 M cacd. | 13.7 | | 177 |
| | 245 | 7.0 | 0.1 M NaCl, 0.05 M cacd., 0.01 M Mg^{++} | 20 | | 177 |

C_8

| | 245 | 7.0 | 0.1 M NaCl, 0.05 M cacd. | 20.2 | | 177 |
| | 245 | 7.0 | 0.1 M NaCl, 0.05 M cacd., 0.01 M Mg^{++} | 27.1 | | 177 |

C_9

| | 245 | 7.0 | 0.1 M NaCl, 0.05 M cacd. | 28.8 | | 177 |
| | 245 | 7.0 | 0.1 M NaCl, 0.05 M cacd., 0.01 M Mg^{++} | 37.7 | | 177 |

Complex	Method	pH	Medium	T_m $^{\circ}$C	Note	Ref.
Poly I Complexes (Con't)						
C_{10}	245	7.0	0.1 $\underline{M}$ NaCl, 0.05 $\underline{M}$ cacd.	36.3		177
	245	7.0	0.1 $\underline{M}$ NaCl, 0.05 $\underline{M}$ cacd., 0.01 $\underline{M}$ Mg^{++}	46.5		177
C_{12}	245	7.0	0.1 $\underline{M}$ NaCl, 0.05 $\underline{M}$ cacd.	48		177
	245	7.0	0.1 $\underline{M}$ NaCl, 0.05 $\underline{M}$ cacd., 0.01 $\underline{M}$ Mg^{++}	57.8		177
Poly dT Complexes						
$d(pA)_{16}$	OD	7.0	0.04 $\underline{M}$ K phosp.	33.8	0.10[b]	34
$d(pA)_{17}$	OD	7.0	0.04 $\underline{M}$ K phosp.	34.3	0.10[b]	34
$d(pA)_{18}$	OD	7.0	0.04 $\underline{M}$ K phosp.	35.8	0.10[b]	34
	OD	N	SSC	46.0	0.08[b]	34
$d(pA)_{19}$	OD	N	SSC	46.6	0.09[b]	34
$d(pA)_{20}$	OD	7.0	0.04 $\underline{M}$ K phosp.	38.0	0.10[b]	34
	OD	N	SSC	47.5	0.08[b]	34
$d(pA)_{21}$	OD	7.0	0.04 $\underline{M}$ K phosp.	38.3	0.10[b]	34
$d(pA)_{22}$	OD	7.0	0.04 $\underline{M}$ K phosp.	40.0	0.10[b]	34
$d(pA)_{25}$	OD	7.0	0.04 $\underline{M}$ K phosp.	41.0	0.10[b]	34
$d(pA)_n$	OD	7	SSC or 0.04 $\underline{M}$ K phosp.	VAR	VAR n	35
	OD	7	0.04 $\underline{M}$ K phosp., 8 m$\underline{M}$ MgCl$_2$	VAR	VAR n	35
2 Poly dT Complexes						
$d(pA)_4$	OD	7.0	0.04 $\underline{M}$ K phosp., 8 $\underline{M}$ mgCl$_2$	16.8	0.14[b]	34
$d(pA)_5$	OD	7.0	0.04 $\underline{M}$ K phosp., 8 $\underline{M}$ MgCl$_2$	25.5	0.14[b]	34
$d(pA)_6$	OD	7.0	0.04 $\underline{M}$ K phosp.	9.5	0.10[b]	34

Complex	Method	pH	Medium	T $^\circ$C	Note	Ref.

2 Poly dT Complexes (Con't)

$d(pA)_6$ (Con't)

Complex	Method	pH	Medium	T $^\circ$C	Note	Ref.
	OD	7.0	0.04 $\underline{M}$ K phosp.;8 m$\underline{M}$ MgCl$_2$	30.4	0.14[b]	34
	OD	N	SSC	19.5	0.11[b]	34
$d(pA)_7$	OD	7.0	0.04 $\underline{M}$ K phosp.	14.0	0.10[b]	34
	OD	7.0	0.04 $\underline{M}$ K phosp.,8 m$\underline{M}$ MgCl$_2$	35.7	0.20[b]	34
	OD	N	SSC	24.0	0.16[b]	34
$d(pA)_8$	OD	7.0	0.04 $\underline{M}$ K phosp.,8 m$\underline{M}$ MgCl$_2$	38.6	0.20[b]	34
	OD	N	SSC	27.5	0.16[b]	34
	OD	N	SSC	VAR	VAR concn. of $d(pA)_8$	34
$d(pA)_9$	OD	7.0	0.04 $\underline{M}$ K phosp.,8 m$\underline{M}$ MgCl$_2$	41.5	0.20[b]	34
	OD	N	SSC	30.2	0.14[b]	34
$d(pA)_{10}$	OD	7.0	0.04 $\underline{M}$ K phosp.	20.0	0.11[b]	34
	OD	N	SSC	34.0	0.15[b]	34
	OD	N	SSC	VAR	VAR concn. of $d(pA)_{10}$	34
$d(pA)_{11}$	OD	N	SSC	35.4	0.16[b]	34
$d(pA)_{17}$	OD	7.0	0.04 $\underline{M}$ K phosp.,8 m$\underline{M}$ MgCl$_2$	53.5	0.30[c]	34
$d(pA)_{21}$	OD	7.0	0.04 $\underline{M}$ K phosp.,8 m$\underline{M}$ MgCl$_2$	55.0	0.30[c]	34
$d(pA)_{25}$	OD	7.0	0.04 $\underline{M}$ K phosp.,8 m$\underline{M}$ MgCl$_2$	56.6	0.30[c]	34
$d(pA)_n$	OD	7.0	0.04 $\underline{M}$ K phosp.	VAR	VAR n	34
	OD	7.0	0.04 $\underline{M}$ K phosp.,8 m$\underline{M}$ MgCl$_2$	VAR	VAR n	34
	OD	N	SSC	VAR	VAR n	34

Poly U Complexes

Complex	Method	pH	Medium	T $^\circ$C	Note	Ref.
Ado	ORD350	7	0.4 $\underline{M}$ NaCl,1 m$\underline{M}$ EDTA,phosp.	VAR	VAR U/Ado	99,268

Complex	Method	pH	Medium	T_m °C	Note	Ref.
Poly U Complexes (Con't)						
n^2m^6Ado	IR	7.4	VAR [Na$^+$]; phosp.	VAR	$2 \rightarrow 1$	102
	IR	7.4	0.51 $\underline{M}$ Na$^+$	29.5	0.06 $\underline{M}$ U	102
(Ap)$_3$A	OD			19	hper. 36%	274
(Ap)$_4$A	OD			28	hper. 39%	274
	OD			VAR	VAR concn. of (Ap)$_4$A	274
(Ap)$_4$	OD			11.5	hper. 34%	274
(Ap)$_5$	OD			20.1	hper. 37%	274
(Ap)$_4$G	OD			13.2	hper. 31%	274
(Ap)$_5$G	OD			21.2	hper. 35%	274
(Ap)$_4$U	OD			12.1	hper. 32%	274
(Ap)$_5$U	OD			20.1	hper. 36%	274
(pA)$_4$U	OD	7.4	1 $\underline{M}$ NaCl,2 m$\underline{M}$ K phosp., 1 m$\underline{M}$ MgCl$_2$	16.3	d.s.cx.	147
(pA)$_4$U$_p$	OD	7.4	1 $\underline{M}$ NaCl,2 m$\underline{M}$ K phosp., 1 m$\underline{M}$ MgCl$_2$	11.5	d.s.cx.	147
G(pA)$_4$	OD			14.4	hper. 36%	274
G(pA)$_5$	OD			23.0	hper. 37%	274
Nuc	ORD350	7	0.4 $\underline{M}$ NaCl,1 m$\underline{M}$ EDTA,phosp.	VAR	d	99,268
U(pA)$_4$	OD			15.2	hper. 35%	274
2 Poly U Complexes						
Ade	1623,1657		0.15 $\underline{M}$ Na$^+$	(23)	0.1 $\underline{M}$ U, 0.067 $\underline{M}$ Ade	98

Complex	Method	pH	Medium	T_m °C	Note	Ref.
2 Poly U Complexes (Con't)						
n^2Ade	1614,1657		0.15 $\underline{M}$ Na$^+$	(43)	0.09 $\underline{M}$ U, 0.045 $\underline{M}$ nAde	98
$m^{6,9}$Ade	1423		H$_2$O	(16)		182
	1657		H$_2$O	(14)		182
Ado	CAL	6.8	0.6 $\underline{M}$ NaCl,0.01 $\underline{M}$ cacd.	32.7		224
	ORD350	7.5	0.4 $\underline{M}$ NaCl,0.01 $\underline{M}$ Tris	15		192
	1624		0.15 $\underline{M}$ Na$^+$	18.5	0.30 $\underline{M}$ U, 0.015 $\underline{M}$ A;co.	98
	1624		0.15 $\underline{M}$ Na$^+$	VAR	2:1 mixt; VAR concn.	98,182
n^2Ado	CAL	6.8	0.6 $\underline{M}$ NaCl,0.01 $\underline{M}$ Na cacd.	49.5		224
	IR		VAR [Na$^+$]	VAR	2:1 complex	182
	1613		0.15 $\underline{M}$ Na$^+$	VAR	2:1 mixt.; VAR concn.	98,182
	1613		0.15 $\underline{M}$ Na$^+$	31.0	0.30 $\underline{M}$ U, 0.015 $\underline{M}$ nAdo;co.	98
	1613		VAR [Na$^+$]	VAR	0.30 $\underline{M}$ Poly U, 0.015 $\underline{M}$ nAdo	98
mn^2Ado	CAL	6.8	0.6 $\underline{M}$ NaCl,0.01 $\underline{M}$ Na cacd.	(39.4)		224
n^2m^6Ado	IR	7.4	VAR [Na$^+$]; phosp.	VAR	3 → 1	102
	IR	(7.4)	0.15 $\underline{M}$ Na$^+$	24.5		102
A-A	760	7.0	0.1 $\underline{M}$ NaCl,0.05 $\underline{M}$ Na cacd., 0.01 $\underline{M}$ Mg^{++}	13.5		177
	OD	7.4	1 $\underline{M}$ NaCl,2 m$\underline{M}$ phosp., 1 m$\underline{M}$ MgCl$_2$	6.2	U/A = 2	147
	260	7.5	0.01 $\underline{M}$ MgCl$_2$,0.01 $\underline{M}$ Tris	13.6		259
	CD260	7.5	0.01 $\underline{M}$ MgCl$_2$,0.01 $\underline{M}$ Tris	13.8		259
A2'-5'A	260	7.0	0.1 $\underline{M}$ NaCl,0.05 $\underline{M}$ Na cacd., 0.01 $\underline{M}$ Mg^{++}	10.6		177
	260	7.5	0.01 $\underline{M}$ MgCl$_2$,0.01 $\underline{M}$ Tris	11.3		259

Complex	Method	pH	Medium	T_m °C	Note	Ref.
2 Poly U Complexes (Con't)						
A2'-5'A (Con't)						
	CD260	7.5	0.01 $\underline{M}$ MgCl$_2$,0.01 $\underline{M}$ Tris	11		259
A-Ap	OD	7.4	1 $\underline{M}$ NaCl,2 m$\underline{M}$ phosp., 1 m$\underline{M}$ MgCl$_2$	5	U/A = 2	147
pA-A	OD	7.4	1 $\underline{M}$ NaCl,2 m$\underline{M}$ phosp., 1 m$\underline{M}$ MgCl$_2$	7	U/A = 2	147
$_L$(A-A)	260	7.5	0.01 $\underline{M}$ MgCl$_2$,0.01 $\underline{M}$ Tris	5.6		259
	CD260	7.5	0.01 $\underline{M}$ MgCl$_2$,0.01 $\underline{M}$ Tris	5.8		259
$_L$(A2'-5'A)	260	7.5	0.01 $\underline{M}$ MgCl$_2$,0.01 $\underline{M}$ Tris	13.5		259
	CD260	7.5	0.01 $\underline{M}$ MgCl$_2$,0.01 $\underline{M}$ Tris	13		259
(Ap)$_2$A	260	7.0	0.1 $\underline{M}$ NaCl,0.05 $\underline{M}$ Na cacd., 0.01 $\underline{M}$ Mg^{++}	25.7		177
	OD	7.4	1 $\underline{M}$ NaCl,2 m$\underline{M}$ phosp., 1 m$\underline{M}$ MgCl$_2$	17.4	U/A = 2	147
(2'-5')(Ap)$_2$A		7.0	0.1 $\underline{M}$ NaCl,0.05 $\underline{M}$ Na cacd., 0.01 $\underline{M}$ Mg^{++}	20.5		177
	260					
(Ap)$_3$	OD	7.4	1 $\underline{M}$ NaCl,2 m$\underline{M}$ phosp., 1 m$\underline{M}$ MgCl$_2$	23	U/A = 2	147
(2'-5')(Ap)$_3$		7.4	1 $\underline{M}$ NaCl,2 m$\underline{M}$ phosp., 1 m$\underline{M}$ MgCl$_2$	13.7	U/A = 2	147
	OD					
(pA)$_3$	OD	6.9	0.15 $\underline{M}$ NaCl,0.045 $\underline{M}$ Na cacd. 0.05 $\underline{M}$ MgCl$_2$	17.6	U/A = 2	147
	OD	7.4	1 $\underline{M}$ NaCl,2 m$\underline{M}$ phosp., 1 m$\underline{M}$ MgCl$_2$	17	U/A = 2	147
d(pA)$_3$	OD	7.4	1 $\underline{M}$ NaCl,2 m$\underline{M}$ phosp., 1 m$\underline{M}$ MgCl$_2$	23	U/A = 2	147
(Ap)$_3$A	260	7.0	0.1 $\underline{M}$ NaCl,0.05 $\underline{M}$ Na cacd., 0.01 $\underline{M}$ Mg^{++}	32.6		177

Complex	Method	pH	Medium	T_m °C	Note	Ref.
2 Poly U Complexes (Con't)						
(Ap)$_3$A (Con't)	260	7.30	0.5 M NaCl,0.04 M phosp.	(26)		29
	OD	7.4	1 M NaCl,2 mM phosp., 1 mM MgCl$_2$	31.3	U/A = 2	147
(2'-5')(Ap)$_3$A 260		7.0	0.1 M NaCl,0.05 M na cacd., 0.01 M Mg^{++}	28.0		177
(Ap)$_4$	OD	7.4	1 M NaCl,2 mM phosp., 1 mM MgCl$_2$	33.2	U/A = 2	147
(pA)$_4$	OD	6.9	0.15 M NaCl,0.045 M Na cacd., 0.05 M MgCl$_2$	32.2	U/A = 2	147
	OD	7.4	1 M NaCl,2 mM phosp., 1 mM MgCl$_2$	29.5	U/A = 2	147
		7.4	VAR [NaCl],0.8 mM MgCl$_2$	VAR	U/A = 2	147
	OD		1 M MgCl$_2$	25	U/A = 2	147
	OD		4 mM MgCl$_2$	31.5	U/A = 2	147
(Ap)$_4$A	260	7.0	0.1 M NaCl,0.05 M Na cacd., 0.01 M Mg^{++}	40.2		177
	257,280	7.5	1.5 mM MnCl$_2$,0.1 M Tris	37	U/A = 2	245
(2'-5')(Ap)$_4$A 260		7.0	0.1 M NaCl,0.05 M Na cacd., 0.01 M Mg^{++}	33.9		177
(Ap)$_5$	OD	7.4	1 M NaCl,2 mM phosp., 1 mM MgCl$_2$	40.5	U/A = 2	147
(pA)$_5$	OD	7.4	1 M NaCl,2 mM phosp., 1 mM MgCl$_2$	37	U/A = 2	147
(Ap)$_5$A	260	7.0	0.1 M NaCl,0.05 M Na cacd., 0.01 M Mg^{++}	43.0		177
	OD	7.0	0.1 M NaCl,0.05 M Na cacd., 0.01 M Mg^{++}	42.5		177

Complex	Method	pH	Medium	T_m °C	Note	Ref.

2 Poly U Complexes (Con't)
_____

Complex	Method	pH	Medium	T_m °C	Note	Ref.
$(2'-5')(Ap)_5A$	260	7.0	0.1 $\underline{M}$ NaCl,0.05 $\underline{M}$ Na cacd., 0.01 $\underline{M}$ Mg^{++}	39.3		177
$(Ap)_6A$	259	7.0	0.1 $\underline{M}$ NaCl,0.1 $\underline{M}$ Tris	(27)	$10^{-5}\underline{M}$ N;co.	192
	ORD350	7.0	0.1 $\underline{M}$ NaCl,0.1 $\underline{M}$ Tris	(40)	$10^{-2}\underline{M}$ N;co.	192
	259	7.0	0.4 $\underline{M}$ NaCl,0.01 $\underline{M}$ Tris	39.5	0.08 m$\underline{M}$ U; A/U = 0.05;hper.7%	192
	259	7.0	0.4 $\underline{M}$ NaCl,0.01 $\underline{M}$ Tris	39.8	0.08 m$\underline{M}$ U; A/U = 0.1;hper.12%	192
	259	7.0	0.4 $\underline{M}$ NaCl,0.01 $\underline{M}$ Tris	40.0	0.08 m$\underline{M}$ U; A/U - 0.2;hper.23%	192
	259	7.0	0.4 $\underline{M}$ NaCl,0.01 $\underline{M}$ Tris	40.5	0.05 m$\underline{M}$ U; A/U = 0.5;hper.37%	192
	259	7.0	0.4 $\underline{M}$ NaCl,0.01 $\underline{M}$ Tris	41.0	0.025 m$\underline{M}$ U; A/U = 1;hper.30%	192
	259	7.0	0.4 $\underline{M}$ NaCl,0.01 $\underline{M}$ Tris	42.0	0.025 m$\underline{M}$ U; A/U = 2;hper.23%	192
	259	7.0	0.4 $\underline{M}$ NaCl,0.01 $\underline{M}$ Tris	44.6	0.025 m$\underline{M}$ U; A/U = 2;hper.16%	192
	260	7.0	0.1 $\underline{M}$ NaCl,0.05 $\underline{M}$ Na cacd., 0.01 $\underline{M}$ Mg^{++}	48.1		177
$(pA)_7$	OD	7.4	1 $\underline{M}$ NaCl,2 m$\underline{M}$ phosp., 1 m$\underline{M}$ MgCl$_2$	49.7	U/A = 2	147
$(Ap)_6A \cdot Ado$	ORD350	7.5	0.4 $\underline{M}$ NaCl,0.01 $\underline{M}$ Tris	19.5;52.5		192
$(Ap)_7A$	260	7.0	0.1 $\underline{M}$ NaCl,0.05 $\underline{M}$ Na cacd., 0.01 $\underline{M}$ Mg^{++}	52.8		177
$(2'-5')(Ap)_7A$	260	7.0	0.1 $\underline{M}$ NaCl,0.05 $\underline{M}$ Na cacd., 0.01 $\underline{M}$ Mg^{++}	47.0		177
$(pA)_9$	OD	7.4	1 $\underline{M}$ NaCl,2 m$\underline{M}$ phosp., 1 m$\underline{M}$ MgCl$_2$	54.6	U/A = 2	147
$(Ap)_{11}$	260	7.0	0.1 $\underline{M}$ NaCl,0.05 $\underline{M}$ Na cacd., 0.01 $\underline{M}$ Mg^{++}	64.4		177

Complex	Method	pH	Medium	T_m °C	Note	Ref.
$(Ap)_{29}A$	260	7.0	0.1 M NaCl,0.05 M Na cacd., 0.01 M Mg^{++}	72.0		177
$(Ap)_{44}A$	260	7.0	0.1 M NaCl,0.05 M Na cacd., 0.01 M Mg^{++}	74.2		177
$(Ap)_{59}A$	260	7.0	0.1 M NaCl,0.05 M Na cacd., 0.01 M Mg^{++}	75.2		177
$(Ap)_{199}A$	260	7.0	0.1 M NaCl,0.05 M Na cacd., 0.01 M Mg^{++}	78.1		177
$(pA)_{\sim 200}$	OD	7.4	1 M NaCl,2 mM phosp., 1 mM MgCl$_2$	72	U/A = 2	147
$(Ap)_4C$	260			16.3	hper.33%	274
$(Ap)_5C$	260			22.7	hper.32%	274
$(pA)_3U$	OD	6.9	0.15 M NaCl ,0.05 M MgCl$_2$, 0.045 M Na cacd.	11.4	t.s.cx.	147
$(pA)_4U$	OD	6.9	0.15 M NaCl,0.05 M MgCl$_2$, 0.045 M Na cacd.	20	t.s.cx.;hpo.30%	147
$C(pA)_4$	260			20.2	hper.38%	274
$C(pA)_5$	260			28.6	hper.40%	274

2 Poly fl^{5}U Complexes

Complex	Method	pH	Medium	T_m °C	Note	Ref.
$(Ap)_5A$	OD	7.0	0.1 M NaCl,0.05 M Na cacd., 0.01 M Mg^{++}	35.9		177

2 Poly cl^{5}U Complexes

Complex	Method	pH	Medium	T_m °C	Note	Ref.
$(Ap)_5A$	OD	7.0	0.1 M NaCl,0.05 M Na cacd., 0.01 M Mg^{++}	67.9		177

2 Poly br^{5}U Complexes

Complex	Method	pH	Medium	T_m °C	Note	Ref.
A-A	260	7.0	0.1 M NaCl,0.05 M Na cacd., 0.01 M MgCl$_2$	34.5		177

Complex	Method	pH	Medium	T_m °C	Note	Ref.

2 Poly br^{5}U Complexes (Con't)

Complex	Method	pH	Medium	T_m °C	Note	Ref.
(Ap)$_2$A	260	7.0	0.1 $\underline{M}$ NaCl,0.05 $\underline{M}$ Na cacd., 0.01 $\underline{M}$ MgCl$_2$	48		177
(Ap)$_4$A	260	7.0	0.1 $\underline{M}$ NaCl,0.05 $\underline{M}$ Na cacd., 0.01 $\underline{M}$ MgCl$_2$	59.5		177
(Ap)$_5$A	260	7.0	0.1 $\underline{M}$ NaCl,0.05 $\underline{M}$ Na cacd., 0.01 $\underline{M}$ MgCl$_2$	66.2		177
(Ap)$_6$A	260	7.0	0.1 $\underline{M}$ NaCl,0.05 $\underline{M}$ Na cacd., 0.01 $\underline{M}$ MgCl$_2$	69.6		177
(Ap)$_7$A	260	7.0	0.1 $\underline{M}$ NaCl,0.05 $\underline{M}$ Na cacd., 0.01 $\underline{M}$ MgCl$_2$	73.7		177

2 Poly io^{5}U Complexes

Complex	Method	pH	Medium	T_m °C	Note	Ref.
(Ap)$_5$A	OD	7.0	0.1 $\underline{M}$ NaCl,0.05 $\underline{M}$ Na cacd., 0.01 $\underline{M}$ Mg^{++}	66.6		177

Poly (U,m^{5}U) Complexes

Complex	Method	pH	Medium	T_m °C	Note	Ref.
(Ap)$_{\sim 12}$	(N)	(SSC)		21.5	hper.14%	253

Poly X Complexes

Complex	Method	pH	Medium	T_m °C	Note	Ref.
A-A	250	7.0	0.1 $\underline{M}$ NaCl,0.05 $\underline{M}$ Na cacd.		no complex	177
(Ap)$_2$A	250	7.0	0.1 $\underline{M}$ NaCl,0.05 $\underline{M}$ Na cacd.	32.7		177
(Ap)$_4$A	250	7.0	0.1 $\underline{M}$ NaCl,0.05 $\underline{M}$ Na cacd.	47.3		177
(Ap)$_5$A	250	7.0	0.1 $\underline{M}$ NaCl,0.05 $\underline{M}$ Na cacd.	52.0		177
(Ap)$_6$A	250	7.0	0.1 $\underline{M}$ NaCl,0.05 $\underline{M}$ Na cacd.	56.6		177
(Ap)$_7$A	250	7.0	0.1 $\underline{M}$ NaCl,0.05 $\underline{M}$ Na cacd.	59.3		177
(Ap)$_{11}$A	250	7.0	0.1 $\underline{M}$ NaCl,0.05 $\underline{M}$ Na cacd.	67.3		177
(Ap)$_{29}$A	250	7.0	0.1 $\underline{M}$ NaCl,0.05 $\underline{M}$ Na cacd.	79.4		177

Complex	Method	pH	Medium	T_m °C	Note	Ref.
Poly X Complexes (Con't)						
$(Ap)_{44}A$	250	7.0	0.1 $\underline{M}$ NaCl, 0.05 $\underline{M}$ Na cacd.	82.5		177
2 Poly X Complexes						
$(Ap)_5A$	OD	7.0	0.1 $\underline{M}$ NaCl, 0.05 $\underline{M}$ Na cacd., 0.01 $\underline{M}$ Mg^{++}	47.2		177

FOOTNOTES FOR SECTION 3.2

[a] See the reference for hypochromicities (260 mμ, 0°, 1 $\underline{M}$ NaCl + 0.1 $\underline{M}$ Na phosphate, pH 6.3 and 262.5 mμ, 0.5°, 1 $\underline{M}$ LiCl + 0.01 $\underline{M}$ cacodylate, pH 6.9) of mixtures of poly A and d$(pT)_n$, n = 4-10.

[b] Slope of the transition curve, $(°C)^{-1}$ measured at T_m where it is maximal; it is defined as the reciprocal of the temperature interval during which a transition curve with maximum slope would take place completely.

[c] The first melting phase is extremely time dependent; as much as 8 hours may be required for equilibration.

[d] Nucleosides tested were: adenosine, L-adenosine, deoxyadenosine, cytidine, N-6-methyladenosine, and inosine.

3.3. POLYNUCLEOTIDES

Polymer	Method	pH	Medium	T^o_m C	Note	Ref.
			3.3.1. <u>Homopolymers</u>			
Poly A	OD	4.0	0.15 <u>M</u> Na acet.	>100		180
	260	4.5	0.5 <u>M</u> Na acet.	(~80)		151
	260	5.0	0.5 <u>M</u> Na acet.	(51)		151
	260	5.5	0.5 <u>M</u> Na acet.	(32)		151
	260	6.0	0.5 <u>M</u> Na acet.	(5)		151
	ORD250,290	4.58	0.15 <u>M</u> KCl, acet.	76		92
	260	4.6	0.15 <u>M</u> KCl,0.05 <u>M</u> acet.	79.5	co.;hper.33%	12
	260	4.63	0.08 <u>M</u> NaCl,5 m<u>M</u> Na acet.	(70)		61
	260	5.05	0.08 <u>M</u> NaCl,5 m<u>M</u> Na acet.	(55)		61
	260	5,53	0.08 <u>M</u> NaCl,5 m<u>M</u> Na acet.	(35)		61
	260	>4.7	<0.2 <u>M</u> acet.	62.3	1.9^j	149a
	260	>4.7	<0.2 <u>M</u> acet.; nC_2n	66.6	7.4^j	149a
	260	>4.7	<0.2 <u>M</u> acet.; nC_3n	66.7	6.0^j	149a
	260	>4.7	<0.2 <u>M</u> acet.; nC_5n	60.8	5.0^j	149a
	260	>4.7	<0.2 <u>M</u> acet.; nC_6n	60.5	6.4^j	149a
	260	>4.7	<0.2 <u>M</u> acet.; nC_7n	57.6	7.8^j	149a
	260	>4.7	<0.2 <u>M</u> acet.; nC_8n	57.1	6.8^j	149a
	OD;ORD	4.85	0.1 <u>M</u> NaCl,0.1 <u>M</u> Na acet.	62		269
	ORD282	4.85	0.1 <u>M</u> NaCl,0.1 <u>M</u> Na acet.	(~60)	co.	215
	ORD436	4.85	0.2 <u>M</u> Na acet.	(61)		270
	ORD	4.85	VAR conductivity; Na acet.	VAR		270
	257	4.86	acet.	(68)	co.	61
	260	5.0	0.03 <u>M</u> Na^+, Na acet.	(78)		151
	260	5.0	0.635 <u>M</u> Na^+, Na acet.	(70)		151
	260	5.0	0.097 <u>M</u> Na^+, Na acet.	(66)		151
	260	5.0	0.181 <u>M</u>, Na^+, Na acet.	(59)		151

Footnotes for Section 3.3 start on page 182.

Polymer	Method	pH	Medium	T_m °C	Note	Ref.
Poly A (Con't)						
	260	5.0	0.265 M Na+, Na acet.	(56)		151
	260	5.0	0.1 M Na+, acet.	66.2±0.2	a	64
	260	5.0	0.1 M Na acet.	63.2	hper.25%;co.	118
	260	5.0	VAR[Na+]	VAR		151
		5	VAR[Na+]	VAR	h.p.hx.,d.s.hx.	76
	257	5.05	0.02 M acet.	(79)	commercial ppn.	7
	257	5.05	0.2 M NaCl,0.02 M acet.	(50)		7
	257	5.1	0.02 M acet.	73.7	hper.53%	7
	257	5.1	0.02 M acet.	56.5	sample preheated at 85-90° for 3 hrs.	7
		5.15	0.02 M acet.	(69)	hper.44%	7
	ORD250,290	5.45	0.15 M KCl, acet.	43		92
	ORD	5.45		43		2
	max	5.5	0.05 M KF, 1 mM EDTA	45.4		70
	252.5	6.0	0.1 M phosp., 1 mM EDTA, 1 mM Na cacd.	(22.3)	$S_{20,w}$=8.2[b]	87
			0.1 M phosp., 1 mM EDTA, 1 mM Na cacd.	(22.6)	$S_{20,w}$=19.3[b]	87
	260	6.3	0.15 M KCl,0.01 M cacd.	9.5	hper.14 and 38%[c]	12
	OD	6.95	0.01 M cacd., VAR [NaCl]	VAR		278
	250	7.0	0.1 M NaCl,0.05 M Na cacd.	88.2		177
	260	7.0	0.15 M KCl,0.01 M cacd.		nco.;hper. and 33%[c]	12
	258	7	0.1 M LiCl,0.01 M cacd.		nco.[d]	135
	EB	N	Mg^{++}	(37)	Mg/N=1;b.t.	117
	ORD250,290	7.1	0.15 M KCl, phosp.		nco.	92
	257	7.2	5 mM phosp., I = 0.01 M	(36)		7
	257	7.2	0.015 M NaCl, 5 mM phosp.; I = 0.025 M	(38)		7
	257	7.2	0.15 M NaCl, 5 mM phosp.; I = 0.16 M	(38)		7

Polymer	Method	pH	Medium	T_m °C	Note	Ref.
Poly A (Con't)						
	260	7.2	0.15 $\underline{M}$ NaCl,0.01 $\underline{M}$ MgCl$_2$, 0.015 $\underline{M}$ Tris	(55)	nco.	125,126
		7.4	0.02 $\underline{M}$ acet.	(40)	nco.;hper.33%	7
	ORD282	7.5	0.15 $\underline{M}$ KF		nco.	215
	OD;ORD		0.15 $\underline{M}$ KF	(41)	nco.	82
	max	8.8	1 m$\underline{M}$ Tris	44.1		70
	OD	e		75		104
	max	Ac		96.1	acid form^z	149
	260	VAR	0.03,0.15 and 0.5 $\underline{M}$ Na$^+$	VAR		151
	260	VAR	0.15 $\underline{M}$ KCl	VAR		21
	OD	VAR	0.15 $\underline{M}$ KCl	VAR		91
	ORD	VAR	0.2 $\underline{M}$ Na$^+$, acet.	VAR		270
	260	VAR		VAR		45
	OD	VAR	VAR I	VAR		78
Poly m^{1}A	OD	4.0	0.15 $\underline{M}$ Na acet.	(83)		180
Poly m^{6}A	OD	4.8	0.1 $\underline{M}$ NaCl,0.1 $\underline{M}$ acet.	f	nco.	23
	OD	7.5	0.1 NaCl,0.05 $\underline{M}$ Tris	f	nco.	23
Poly m^{6,6}A	OD	4.8	0.1 $\underline{M}$ NaCl,0.1 $\underline{M}$ acet.	f	nco.	23
	OD	7.5	0.1 $\underline{M}$ NaCl,0.05 $\underline{M}$ Tris	f	nco.	23
Poly n^{2}m^{6}A	280	5.3	0.02 $\underline{M}$ Na$^+$, acet.	42		102
	1661,1619	5.3	0.12 $\underline{M}$ Na$^+$, cacd.	14		102
	250,277	7.9	0.1 $\underline{M}$ Na$^+$, cacd.	(50)	nco.	102
	265	7.9	0.1 $\underline{M}$ Na$^+$, cacd.	(55)	nco.	102
	290	7.9	0.1 $\underline{M}$ Na$^+$, cacd.	(57)	nco.	102
Poly Am	260	4.6	0.15 $\underline{M}$ KCl,0.05 $\underline{M}$ acet.	86	14-15S;co.; hper.35%	12

Polymer	Method	pH	Medium	T_m °C	Note	Ref.
Poly Am (Con't)						
	260	5.7	0.15 $\underline{M}$ KCl,0.01 $\underline{M}$ cacd.	47	14–15A;hper.26 and 30%(pH 5.4)[c]	12
	260	6.3	0.15 $\underline{M}$ KCl,0.01 $\underline{M}$ cacd.	32	14–15S;hper.22 and 39%[c]	12
	260	7.0	0.15 $\underline{M}$ KCl,0.01 $\underline{M}$ cacd.	13.5	14–15S;hper.12 and 34%[c]	12
	260	VAR	0.15 $\underline{M}$ KCl	VAR		12
Poly Aac	260	7.2	0.15 $\underline{M}$ NaCl,0.015 $\underline{M}$ Tris 0.01 $\underline{M}$ MgCl$_2$		88% acetylation; nco.;red.hper. upon acetylation 125,126,127	
Poly isoA	OD	7.0	0.02 $\underline{M}$ Na$^+$		no t.	178
Poly c^{7}A	270	7.0	0.1 $\underline{M}$ NaCl,0.05 Na cacd.		nco.	103
Poly dA	ORD	4.3	acet.	65	part.rev.; hper.14%	2
	260	7.75–7.80	VAR[Na$^+$]; 0.1 $\underline{M}$ phosp.	VAR	nco.;rev.	210
Poly C	278	3.3	0.1 $\underline{M}$ NaCl,0.01 $\underline{M}$ acet.	41	co.	22
	278	3.9	0.1 $\underline{M}$ NaCl,0.01 $\underline{M}$ acet.	75	co.	22
	278	4.2	0.1 $\underline{M}$ NaCl,0.01 $\underline{M}$ acet.	77	co.	22
	278	4.45	0.1 $\underline{M}$ NaCl,0.01 $\underline{M}$ acet.	78	co.	22
	278	4.7	0.1 $\underline{M}$ NaCl,0.01 $\underline{M}$ acet.	74	co.	22
	278	4.9	0.1 $\underline{M}$ NaCl,0.01 $\underline{M}$ acet.	70	co.(nco. above pH 5.8)	22
	274	3.65	0.1 $\underline{M}$ Na acet.	73		3
	274	4.05	0.1 $\underline{M}$ Na acet.	81.3		3
	274	4.85	0.1 $\underline{M}$ Na acet.	71.5		3
	OD	3.65	0.1 $\underline{M}$ Na acet.,3 $\underline{M}$ urea	39.5		3
	OD	3.65	0.1 $\underline{M}$ Na acet.,5 $\underline{M}$ urea	<15		3
	OD	4.05	0.1 $\underline{M}$ Na acet.,3 $\underline{M}$ urea	63		3
	OD	4.05	0.1 $\underline{M}$ Na acet.,5 $\underline{M}$ urea	44		3

Polymer	Method	pH	Medium	T_m °C	Note	Ref.
Poly C (Con't)						
	OD	4.85	0.1 $\underline{M}$ Na acet.,3 $\underline{M}$ urea	63		3
	OD	4.85	0.1 $\underline{M}$ Na acet.,5 $\underline{M}$ urea	53.5		3
	278	3.95	0.1 $\underline{M}$ NaCl,0.01 $\underline{M}$ acet.	(65)	co.	71
		4.0	0.1 $\underline{M}$ buff.	82		249
	max	4.0	0.1 $\underline{M}$ NaCl,0.05 $\underline{M}$ acet.	(81)	co.;hper.~20%	204
	OD	4.0	0.1 $\underline{M}$ Na acet.	(66)		269
	ORD	4.0	0.1 $\underline{M}$ Na acet.	(74)		269
	280	4.0	0.1 $\underline{M}$ NaCl,0.05 $\underline{M}$ Na acet.	82.5		177
	280	4	0.1 $\underline{M}$ NaCl,0.05 $\underline{M}$ Na acet.	79.5		175
	260,275	4	Na$^+$; I = 0.13 $\underline{M}$	(74)		77
	276	4.02	0.04 $\underline{M}$ NaClO$_4$	78		285
	276	4.02	0.04 $\underline{M}$ NaClO$_4$ + Cu^{++}	58	10 Cu^{++}/P	285
	275	4.05	0.1 $\underline{M}$ NaCl,0.01 $\underline{M}$ acet.	(84)		120
	260	4.08	0.1 $\underline{M}$ Na acet.	81	co.;rev.; hper.27%	54
	276	4.47	0.04 $\underline{M}$ NaClO$_4$	80.5		285
	276	4.47	0.04 $\underline{M}$ NaCl$_4$ + Cu^{++}	63	10 Cu^{++}/P	285
	280	4.5	0.1 $\underline{M}$ NaCl,0.05 $\underline{M}$ Na acet.	78		175
	273.5	4.5	0.1 $\underline{M}$ NaCl,0.16 m$\underline{M}$ cacd.	73		185
	ORD	4.6	Na acet.; I = 0.2 $\underline{M}$	(75)		213
	275	4.6	Na form; I = 0.05 $\underline{M}$	78		67
	OD;ORD	4.85	0.1 $\underline{M}$ NaCl,0.1 $\underline{M}$ Na acet.	(67)		269
	OD	5.0	0.15 $\underline{M}$ Na$^+$	65.5		180
	280	5	0.1 $\underline{M}$ NaCl,0.05 $\underline{M}$ Na acet.	63.3		175
	260,275	5	Na$^+$;I = 0.13 $\underline{M}$	(69)		77
		5	VAR[Na$^+$]	VAR		76
	max	5.5	0.05 $\underline{M}$ KF, 1 m$\underline{M}$ EDTA	56.1		70
	OD	5.8-6.0	VAR[Cu^{++}]		VAR hper.	285
	OD	5.8-6.0	0.5 m$\underline{M}$ Cu^{++}, VAR [Na^{++}]		VAR hper.	285

Polymer	Method	pH	Medium	T_m °C	Note	Ref.
Poly C (Con't)						
	260	7.0	SSC	(~45)	nco.;rev.; hper.9%8	54
	268	7.0	0.1 $\underline{M}$ NaCl,0.01 $\underline{M}$ Na cacd.		nco.; 85% rev.	214
	278	7	0.1 $\underline{M}$ Na phosp.		nco.;hper.28% (20–90°)	71
	max	7.3	0.01 $\underline{M}$ KF,1 m$\underline{M}$ Tris	48.5	b.t.^{z}	149
		7.8	0.1 $\underline{M}$ NaCl,0.05 $\underline{M}$ Tris		no t.	257
	max	8.8	1 m$\underline{M}$ Tris	49.9		70
	OD	VAR	0.01 $\underline{M}$ Na$^+$	VAR	c·c$^+$	216
	OD	VAR	0.25 $\underline{M}$ Na$^+$	VAR	c·c$^+$	216
	274	VAR	0.1 $\underline{M}$ Na acet.	VAR		3
	274	3.65;4.05 4.85	VAR conductivity; acet.	VAR		3
	OD	VAR	0.15 $\underline{M}$ and 1.0 $\underline{M}$ NaCl	VAR		80
	OD	VAR	VAR[Na$^+$]	VAR		77
	270	VAR	0.5 m$\underline{M}$ Cu^{++}		VAR hper.	285
Poly m⁴C	274	3.9	0.1 $\underline{M}$ Na acet.		nco.;hper.19% (10–60°)	71
	274	7.1	0.1 $\underline{M}$ Na phosp.		nco.;hper.32% (10–90°)	
	274	8.0	0.1 $\underline{M}$ phosp.	(~65)	nco.	22
Poly m⁵C		4.0	0.1 $\underline{M}$ buff.	79		249
	OD	5.8–6.0	VAR [Cu^{++}]		VAR hper.	285
	OD	5.8	0.5 m$\underline{M}$ Cu^{++},VAR[Na$^+$]		VAR hper.	285
		7.8	0.1 $\underline{M}$ NaCl,0.05 $\underline{M}$ Tris		no t.	257
Poly m⁴,⁴C	280	4.11	0.1 $\underline{M}$ Na acet.		no t.	71
	280	7.09	0.1 $\underline{M}$ Na phosp.		nco.; hper.37% (10–90°)	71
	280	8.0	0.1 $\underline{M}$ phosp.	(~60)	nco.	22

Polymer	Method	pH	Medium	T_m °C	Note	Ref.
Poly m^{4,5}C	max	4.0	0.1 M NaCl,0.05 M acet.		nco.;hper.16% (20–85°)	204
	OD	5.8	VAR[Cu^{++}]		VAR hper.	285
	OD	5.8	0.5 mM Cu^{++},VAR [Na$^+$]		VAR hper.	285
	OD	5.8–6.0	VAR[Cu^{++}]		VAR hper.	285
	OD	5.8	0.5 mM Cu^{++},VAR[Na$^+$]			285
	278	7.5	0.1 M NaCl,0.05 M Tris.		nco.;hper.14% (20–80°)	204
Poly br^{5}C	295	3.0	0.1 M NaCl,0.05 M Na acet.	>82		175
	295	3.5	0.1 M NaCl,0.05 M Na acet.	65.2		175
	295	4	0.1 M NaCl,0.05 M Na acet.	45.6		175
	295	4.5	0.1 M NaCl,0.05 M Na acet.	19.5		175
	295	5	0.1 M NaCl,0.05 M Na acet.	<0		175
	255,315	4.0	0.03 M Na$^+$, acet.	57.3		177
	255,315	4.0	0.15 M Na$^+$, acet.	46.6		177
	255,315	4.0	0.30 M Na$^+$, acet.	41.3		177
	255,315	4.0	1.00 M Na$^+$, acet.	37.5		177
	255,315	4.0	1.5 M Na$^+$, acet.	35.4		177
Poly io^{5}C	290	3.0	0.1 M NaCl,0.05 M Na acet.	50.3;94.4		177
	290	3.5	0.1 M NaCl,0.05 M Na acet.	52.8;90.7		177
	290	4.0	0.1 M NaCl,0.05 M Na acet.	49.5;76.5		177
	325	4.0	0.1 M NaCl,0.05 M Na acet.	76.5		177
	ORD365	4.0	0.1 M NaCl,0.05 M Na acet.	50.1;76.7		177
	290	4.5	0.1 M NaCl,0.05 M Na acet.	41.3;54.1		177
	290	5	0.1 M NaCl,0.05 M Na acet.	29.2; ~27		177

Polymer	Method	pH	Medium	T_m °C	Note	Ref.
Poly io^{5}C (Con't)						
	290	4.0	0.045 $\underline{M}$ Na$^+$, acet.	45.8;78.7		177
	290	4.0	0.088 $\underline{M}$ Na$^+$,acet.	46.7;76.8		177
	290	4.0	0.173 $\underline{M}$ Na$^+$,acet.	49.1;73.5		177
	290	4.0	0.445 $\underline{M}$ Na$^+$,acet.	54.2;67.5		177
	290	4.0	0.900 $\underline{M}$ Na$^+$,acet.	61.2		177
Poly ho^{4}C	OD	7	0.1 $\underline{M}$ NaCl		no hper.(15-80°)	119
Poly ho^{4,6}C	OD	7	0.1 $\underline{M}$ NaCl		no hper.(15-80°)	119
Poly sca^{4}C	OD	7	0.1 $\underline{M}$ NaCl		no hper.(15-80°)	119
Poly Cm	245	4.5	0.1 $\underline{M}$ Na$^+$,0.05 $\underline{M}$ acet.	~70	co.;hper.75%	288
	270	4.5	0.1 $\underline{M}$ Na$^+$,0.05 $\underline{M}$ acet.	~70	co.;hper.10%	288
	270	6.5	0.1 $\underline{M}$ Na$^+$,0.05 $\underline{M}$ acet.		nco.;b.t.;hper.15% (22-85°)	288
Poly dC	OD	6.0±0.1	0.1 $\underline{M}$ NaCl	88	d.s.hx.	106
		6.0±0.1	0.3 $\underline{M}$ NaCl	69	d.s.hx.	106
	246	6.1	0.286 $\underline{M}$ Na$^+$	68-70	d.s.hx.	106
	240	6.57	0.42 $\underline{M}$ Na$^+$	(51)	d.s.hx.	106
	270	6.9	0.41 $\underline{M}$ Na$^+$	(45)	d.s.hx.	106
	max	7.3	0.01 $\underline{M}$ KF,1 m$\underline{M}$ Tris		no t.(20-80°)^z	149
	OD	VAR	0.1 and 0.4 $\underline{M}$ Na$^+$	VAR		106
	max	VAR	0.4 $\underline{M}$ Na$^+$	VAR		287
Poly d(m^{5}C)	240	6.52	0.41 $\underline{M}$ Na$^+$	(45)		287
	max	N			hper.5%(20-80°)	287
	OD	VAR	0.4 $\underline{M}$ Na$^+$	VAR		287
Poly d(m^{4,5}C)	max	N			hper.11%(20-80°)	287

Polymer	Method	pH	Medium	T_m °C	Note	Ref.
Poly d(e^{4}m^{5}C)	max	N			hper.2%(20-80°)	287
Poly d(br^{5}C)	285	4.75	0.41 M Na$^+$	(33)		106
Poly G	275	6	H$_2$O	(~60)		92
	OD	7	2 mM Na$^+$		t.incompl.100°	62
	275	(N)	0.01 mM NaCl	73.2		70
	275	(N)	0.1 mM NaCl	75.8		70
	275	(N)	1 mM NaCl	75.2		70
	OD	(N)	2 mM Na$^+$		no hper. < 100°	92
	OD	(N)	0.1 M NaCl,0.05 M cacd.	>100		198
	260	10.0	0.01 M NaCl,5 mM Na bor.; 25% p.g.	84		190
	280	10.0	0.01 M NaCl,5 mM Na bor.; 25% p.g.	88		190
Poly m^{1}G	255	4.0	0.055 M NaCl,2 mM NH$_4$ acet.; 50% e.g.	82.5		196
	260	7.0	0.5 M LiCl,2 mM Na cacd.; 28% su.	(42)		136
Poly m^{7}G	OD	7.0	0.15 M Na$^+$		t.incompl.100°	180
	OD		0.1 M NaCl,0.05 M Na cacd.	>100		198
Poly m^{2,2}G	255	3.0	0.1 M NaCl,0.05 M Na acet.	(43)		198
	255	7.0	H$_2$O	~20		198
	255	7.0	0.1 M NaCl,5 mM Na cacd.	(30)		198
	255	7.0	0.1 M NaCl,0.05 M Na cacd.	47.5		198
Poly br^{8}G		7.0	0.1 M NaCl,0.05 M Na cacd.		f	179
Poly I	250	3.0	Na form.; I = 0.05 M		no hper.<90°	67
	245	4	0.1 M NaCl,0.05 Na acet.	24.8		175
	275	4.6	Na form.; I = 0.05 M		no hper. < 95°	67
	245	5	0.1 M NaCl,0.05 Na acet.	26.2		175

Polymer	Method	pH	Medium	T_m °C	Note	Ref.
Poly I (Con't)						
	max	6	1.0 $\underline{M}$ NaCl,2 m$\underline{M}$ phosp.	(42)	co.;t.s.hx.	50
	248	6.8-7.0	0.389 $\underline{M}$ Na$^+$,0.01 $\underline{M}$ citr.	(41.3)	b.t.[h]	90
	248	6.8-7.0	0.600 $\underline{M}$ Na$^+$,0.01 $\underline{M}$ citr.	(37.5)	t.s.hx.[h]	90
	248	6.8-7.0	0.750 $\underline{M}$ Na$^+$,0.01 $\underline{M}$ citr.	(41.0)	t.s.hx.[h]	90
	248	6.8-7.0	1.000 $\underline{M}$ Na$^+$,0.01 $\underline{M}$ citr.	(42.8)	t.s.hx.[h]	90
	CAL	6.8-7.0	1.000 $\underline{M}$ Na$^+$,0.01 $\underline{M}$ citr.	43.7	t.s.hx.[h]	90
	248;ORD 265,288	7.0	1 $\underline{M}$ NaCl,0.01 $\underline{M}$ EDTA	(42)	co.	214
	OD	7.0	0.15 $\underline{M}$ Na$^+$	27		178
	OD	7.8	VAR [Na$^+$]; phosp.	VAR		40
	245		0.101 $\underline{M}$ Na$^+$	(21)	rev.;hper.36%	106
	245		0.50 $\underline{M}$ Na$^+$	(38)	rev.;hper.36%	106
	245		1.00 $\underline{M}$ Na$^+$	(46)	rev.;hper.36%	106
	OD	VAR	VAR [Na$^+$]	VAR	t.s.hx.	261
Poly dI	243	(6.1)	0.101 $\underline{M}$ Na$^+$	18	d.s.hx.	106
	246	6.1	0.286 $\underline{M}$ Na$^+$	32	d.s.hx.	106
	OD	(6.1)	VAR [Na$^+$]	VAR	d.s.hx.	106
	245	6.4	0.099 $\underline{M}$ Na$^+$	(18)	d.s.hx.	107
	245	6.4	0.212 $\underline{M}$ Na$^+$	(27)	d.s.hx.	107
	245	6.4	0.561 $\underline{M}$ Na$^+$	(31)	d.s.hx.	107
	OD	7.8	VAR [Na$^+$]; phosp.	VAR		40
Poly T	265	3.0	0.15 $\underline{M}$ Na acet.	25.4	6.1[i]	251
	265	4.0	0.15 $\underline{M}$ Na acet.	25.5	5.5[i]	251
	265	4.2	0.15 $\underline{M}$ Na$^+$,5 m$\underline{M}$ Na acet.	27.2	5.3[i]	251
	265	5.5	0.15 $\underline{M}$ Na acet.	26.2	5.8[i]	251
	265	5.6	0.15 $\underline{M}$ NaCl	27.3	5.5[i]	251
		6.8	0.1 $\underline{M}$ KCl,0.05 $\underline{M}$ Tris	29	rev.	248

Polymer	Method	pH	Medium	T_m^o °C	Note	Ref.
Poly T (Con't)						
	OD	7.0	0.1 $\underline{M}$ NaCl,0.1 $\underline{M}$ MgCl$_2$, 0.05 $\underline{M}$ Na cacd.	(40)		170
	266	7	SSC	29-30	rev.	255
	267	7	SSC	27.5	hper.32%	218
	266	7	0.01 $\underline{M}$ MgCl$_2$	36	rev.;hper.85%	255
		7	0.01 $\underline{M}$ MgCl$_2$	35	concn.10^{-3} to 10^{-5} $\underline{M}$	248
	267	7	0.01 $\underline{M}$ MgCl$_2$,0.05 $\underline{M}$ Tris	32.5	hper.30%	218
	OD	N	0.01 $\underline{M}$ MgCl$_2$	36.5		258
		N		33	hper.50%	257
	265	7.2	SSC	29	hper.50%	265
	260	7.4	VAR [Mg^{++}];0.01 $\underline{M}$ phosp.	VAR	k	250
	265	7.8	0.15 $\underline{M}$ Na$^+$,5 m$\underline{M}$ Na phosp.	27.3	5.3[1]	251
	265	(N)	0.05 $\underline{M}$ NaCl	(18.5)	2.8[1];rev.; hper.50%	251
	265	(N)	0.1 $\underline{M}$ NaCl	(24.1)	rev.;hper.50%	251
	265	(N)	0.15 $\underline{M}$ NaCl	(27.3)	5.5[1];hper.50%	251
	265	(N)	0.2 $\underline{M}$ NaCl	(29.0)	rev.;hper.50%	251
	265	(N)	0.3 $\underline{M}$ NaCl	(32.1)	rev.;hper.50%	251
	265	(N)	0.5 $\underline{M}$ NaCl	(36)	hper.50%	251
	265	(N)	1.0 $\underline{M}$ NaCl	(40)	hper.50%	251
	265	(N)	VAR [NH$_4$+],[Li$^+$],[K$^+$],and [Na+]	VAR		251
			<0.01 $\underline{M}$ Na$^+$	27	hper.50%	265
			0.01 $\underline{M}$ Mg^{++}	33.5	hper.50%	265
	265		5 x 10^{-6} $\underline{M}$ Mg^{++}	(23.2)	rev.	251
	265		10^{-5} $\underline{M}$ Mg^{++}	(25.7)	0.3[1];rev.	251
	265		3 x 10^{-5} $\underline{M}$ Mg^{++}	(28.9)	rev.	251
	265		0.1 m$\underline{M}$ Mg^{++}	(32.4)	0.6[1];rev.	251
	265		1 m$\underline{M}$ Mg^{++}	(34.1)	2[1];rev.	251

Polymer	Method	pH	Medium	T_m $^\circ C$	Note	Ref.
Poly T (Con't)						
	265		0.01 $\underline{M}$ Mg^{++}	(36.0)	rev.	251
	265		0.1 $\underline{M}$ Mg^{++}	(38.1)	3.6^1;rev.	251
	265		VAR [Co^{++}],[Mg^{++}],[Cu^{++}], [Ba^{++}], and [Mn^{++}]	VAR		251
	265		VAR concn. of mono,di and tri-substituted alkylamines	VAR		251
	265		distd. H$_2$O,Mg^{++},or pH 7.4 0.01 $\underline{M}$ phosp.;VAR concn. of nC$_z$n,sperm. and sperd.	VAR	z ≤ 6	251
	OD;ORD289		30% MeOH	20		266
	OD;ORD289		40% MeOH	-4		266
	OD;ORD289		50% MeOh	-12	rev.	266
	OD;ORD289		60% MeOH	-8		266
	OD;ORD289		70% MeOH	-15.5	hysteresis	266
	OD;ORD289		80% MeOH	-18		266
Poly dT	260	6.4	0.01 $\underline{M}$ MgCl$_2$	36		227
Poly U	260	6.0	0.2 $\underline{M}$ Na phosp.	(~5)		208a
	260	7.0	0.1 $\underline{M}$ NaCl,0.05 $\underline{M}$ Na cacd.,0.01 $\underline{M}$ MgCl$_2$	4.9		177
	260	7.0	0.1 $\underline{M}$ NaCl,0.05 $\underline{M}$ cacd., 0.02 $\underline{M}$ Mg^{++}	6.6		184
	OD	7.0	0.1 $\underline{M}$ NaCl,0.1 $\underline{M}$ MgCl$_2$, 0.05 $\underline{M}$ Na cacd.	(6)		170
	260	7.0	0.1 $\underline{M}$ NaCl,0.01 $\underline{M}$ MgCl$_2$, 0.05 $\underline{M}$ Na cacd.	4.9		174
	ORD313	7.0	0.1 $\underline{M}$ NaCl,0.01 $\underline{M}$ MgCl$_2$, 0.05 $\underline{M}$ Na cacd.	(4)		174
	260	7	0.135 $\underline{M}$ NaCl,0.015 $\underline{M}$ Na cacd.	-2.8	extrapol.value	174
	260	7	0.27 $\underline{M}$ NaCl,0.030 $\underline{M}$ Na cacd.	2.2	extrapol.value	174
	260	7	0.9 $\underline{M}$ NaCl,0.1 $\underline{M}$ Na cacd.	5.2		174
	ORD313	7	0.9 $\underline{M}$ NaCl,0.1 $\underline{M}$ Na cacd.	(6)		174

Polymer	Method	pH	Medium	T_m °C	Note	Ref.
Poly U (Con't)						
	OD	7	0.15 $\underline{M}$ NaCl,0.015 $\underline{M}$ Na	8.5	hper.18.5%	218
		7	0.01 $\underline{M}$ MgCl$_2$	8.5		248
	266	7	0.01 $\underline{M}$ MgCl$_2$	8.5	rev.	255
	260	7	1 $\underline{M}$ LiCl,0.05 $\underline{M}$ Na cacd.; 28% su.	4		136
	260	7	1 $\underline{M}$ NaCl,0.05 $\underline{M}$ Na cacd.; 28% su.	5.2		136
	260	7	1 $\underline{M}$ CsCl,0.06 $\underline{M}$ Na cacd.; 28% su.	10.5		136
	max	N	0.2 $\underline{M}$ NaCl	(3)	hper.6%	289
	max	N	0.1 $\underline{M}$ MgCl$_2$	(8)	hper.31%	289
	260	7.4	0.01 $\underline{M}$ phosp.,1 m$\underline{M}$ Mg^{++}	(3)	a	250
	259; ORD365	7.4	0.01 $\underline{M}$ MgCl$_2$	(6)		144
	max	7.4	5 m$\underline{M}$ phosp.,0.25 $\underline{M}$ sperd. per mol. U	(23)	hper.38%	289
	max	7.4	5 m$\underline{M}$ phosp.,0.35 $\underline{M}$ sperd. per mol. U	(39)	hper.38%	289
	260	7.4	0.02 $\underline{M}$ phosp.,0.0135 m$\underline{M}$ sperd.	24.2	0.01 m$\underline{M}$ U; hper.43%;co.	252
	260	7.4	0.02 $\underline{M}$ phosp.,0.01 m$\underline{M}$ sperm.	28.3	0.01 m$\underline{M}$ U; hper.43%;co.	252
		7.4	VAR [Mg^{++}]	VAR		252
		7.4	VAR [nC$_z$n]	VAR	$z \leq 6$	252
	ORD284	7.5	0.15 $\underline{M}$ KF	(~10)		215
		8.0		5.2		184
	CD		0.15 $\underline{M}$ KF	5.2	k	82
	260		0.2 $\underline{M}$ Mg(ClO$_4$)$_2$	8		271
	260		VAR [Mg^{++}]	VAR		251
	260		nC$_z$n	VAR	$z \leq 6$	251
	ORD282		50% MeOH	-32		266
Poly m^{3}U	262	(N)	0.01 - 0.05 $\underline{M}$ MgCl$_2$		no hper.(4-75°)	253

Polymer	Method	pH	Medium	T_m °C	Note	Ref.
Poly e^{5}U		N	0.6 $\underline{M}$ NaCl	~0		226
		N	0.01 $\underline{M}$ Mg^{++}	~0		226
	267	N	0.01 $\underline{M}$ Mg^{++}	~-2		247
	267	N	0.01 $\underline{M}$ Mg^{++}	-2.0		246
	267	N	1 equiv. sperm.	9.2		246
	267	N	1 equiv. sperd.	11.5		246
	267	N	1 equiv. sperm.	9.2	hper. 30%	247
	267	N	1 equiv. sperd.	11.5	hper. 32%	247
Poly fl^{5}U	269	N	0.01 $\underline{M}$ MgCl$_2$	< 5	hper. 7.5% (2.5-5°); no t. (5-75°)	256
Poly cl^{5}U	OD	7.0	0.1 $\underline{M}$ NaCl, 0.1 $\underline{M}$ MgCl$_2$, 0.05 $\underline{M}$ Na cacd.	(< 0)		170
		(7.0)	(0.1 $\underline{M}$ NaCl, 0.05 $\underline{M}$ cacd. 0.02 $\underline{M}$ Mg^{++})	1		184
Poly br^{5}U	279	7.0	0.1 $\underline{M}$ NaCl, 0.05 $\underline{M}$ cacd., 0.02 $\underline{M}$ Mg^{++}	9.4		184
	OD	7.0	0.1 $\underline{M}$ NaCl, 0.1 $\underline{M}$ MgCl$_2$, 0.05 $\underline{M}$ Na cacd.	(9)		170
	260,280	7	0.1 $\underline{M}$ Na$^+$, 0.01 $\underline{M}$ phosp.		no t. < 95°	211
Poly io^{5}U	OD	7.0	0.1 $\underline{M}$ NaCl, 0.1 $\underline{M}$ MgCl$_2$, 0.05 $\underline{M}$ Na cacd.	(20)		170
		(7.0)	(0.1 $\underline{M}$ NaCl, 0.05 $\underline{M}$ cacd., 0.02 $\underline{M}$ Mg.$^{++}$)	21.5		184
Poly s^{4}U	330	7	0.1 $\underline{M}$ KCl, 0.01 $\underline{M}$ MgCl$_2$, 0.05 $\underline{M}$ Na cacd.	8		232
Poly s^{2,4}U	288	7	0.05 $\underline{M}$ Na cacd.	80.5		53b
	340	7	0.05 $\underline{M}$ Na cacd.	81		53b
	OD	N	1 m$\underline{M}$ NaCl	74		53b
Poly Um	max	N	0.1 $\underline{M}$ MgCl$_2$	(25)	hper. 34%	289

Polymer	Method	pH	Medium	T_m $^{\circ}C$	Note	Ref.
Poly Um (Con't)						
	max	N	0.2 $\underline{M}$ NaCl	(14)	$S_{20,W}$(0.01 $\underline{M}$ NaCl)= 9.0;hper.31%	289
	max	7.4	5 m$\underline{M}$ phosp.;0.25 $\underline{M}$ sperd. per mol. Um	(24)	hper.41%	289
	max	7.4	5 m$\underline{M}$ phosp.;0.35 $\underline{M}$ sperd. per mol. Um	(28)	hper.41%	289
Poly dU			0.01 to 0.02 $\underline{M}$ Mg^{++}		no hper. >0°	286
Poly Q	OD	7.0		(60)		170
	OD	7.0	0.1 $\underline{M}$ NaCl,0.01 $\underline{M}$ Mg^{++}, 0.05 $\underline{M}$ cacd.	60	co.	200
	OD	8.0	0.1 $\underline{M}$ NaCl,0.01 $\underline{M}$ Mg^{++}, 0.05 $\underline{M}$ cacd.	58	co.	200
	OD	8.5	0.1 $\underline{M}$ NaCl,0.01 $\underline{M}$ Mg^{++}, 0.05 $\underline{M}$ cacd.	56	co.	200
	OD	9.0	0.1 $\underline{M}$ NaCl,0.01 $\underline{M}$ Mg^{++}, 0.05 $\underline{M}$ cacd.	51	co.	200
	265	7.0	0.1 $\underline{M}$ NaCl,0.01 $\underline{M}$ Mg^{++}, 0.05 $\underline{M}$ cacd.	60	co.	73
	265	8.0	0.1 $\underline{M}$ NaCl,0.01 $\underline{M}$ Mg^{++}, 0.05 $\underline{M}$ cacd.	58	co.	73
	265	8.5	0.1 $\underline{M}$ NaCl,0.01 $\underline{M}$ Mg^{++}, 0.05 $\underline{M}$ cacd.	56	co.	73
	265	9.0	0.1 $\underline{M}$ NaCl,0.01 $\underline{M}$ Mg^{++}, 0.05 $\underline{M}$ cacd.	51	co.	73
Poly X	275	5	0.1 $\underline{M}$ NaCl,0.05 $\underline{M}$ Na acet.	44.5		176
	275	5.8	0.1 $\underline{M}$ NaCl,0.05 $\underline{M}$ Na acet.	43.0		176
	275	6.0	0.1 $\underline{M}$ NaCl,0.05 $\underline{M}$ Na acet.	42.5		176
	275	7.0	0.1 $\underline{M}$ NaCl,0.05 $\underline{M}$ Na cacd.	33.4		176
	275	7.3	0.15 $\underline{M}$ NaCl,7.5 m$\underline{M}$ Tris	28.2		176
	275	7.65	0.15 $\underline{M}$ NaCl,7.5 m$\underline{M}$ Tris	24.5		176
	275	8.32	0.15 $\underline{M}$ NaCl,7.5 m$\underline{M}$ Tris	25.6		176

Polymer	Method	pH	Medium	T_m °C	Note	Ref.
Poly X (Con't)						
	275	8.50	0.15 $\underline{M}$ NaCl,7.5 m$\underline{M}$ Tris	23.5		176
	260	4.7	0.05 $\underline{M}$ Na$^+$		nco.;hper.8%, 17^j	57
	260	5.2	0.1 $\underline{M}$ Na$^+$	(34)	hper.10%;7-8^j	57
	260	5.2	0.1 $\underline{M}$ Na$^+$,0.05 $\underline{M}$ acet.	30	hper.10%;2.5^j	262
	260	5.6	0.05 $\underline{M}$ Na$^+$,0.05 $\underline{M}$ acet.	36	hper.22%;6.5^j	262
	260	5.6	0.1 $\underline{M}$ Na$^+$, 0.05 $\underline{M}$ acet.	38	hper.21%;6^j	262
	260	5.6	0.15 $\underline{M}$ Na$^+$,0.05 $\underline{M}$ acet.	40	hper.20%;6.5^j	262
	260	5.6	0.1 $\underline{M}$ Na$^+$	(38)	hper.22%	57
	260	5.7	VAR [Mg^{++}]; 0.01 $\underline{M}$ acet.	VAR		57
		5.7	0.05 $\underline{M}$ Na^{++}	35		57
		5.7	1.0 $\underline{M}$ Na$^+$	56		57
		6.6	0.05 $\underline{M}$ Na$^+$	35		57
		6.6	0.2 $\underline{M}$ Na$^+$	37.5		57
	260	6.6	0.1 $\underline{M}$ Na$^+$	(37)	co.;hper.40%	57
	275	7.0	0.1 $\underline{M}$ Na$^+$,0.025 $\underline{M}$ phosp.	(33)	hper.40-55%^{l}	57
	250	7.0	0.1 $\underline{M}$ Na$^+$,0.025 $\underline{M}$ phosp.	(32)	hper. ~20%	57
	275	7.0	0.1 $\underline{M}$ Na$^+$,0.025 $\underline{M}$ phosp.	33	hper.50%; 4^j	262
	250	7.0	0.1 $\underline{M}$ NaCl,0.05 $\underline{M}$ Na cacd.	33.4		177
	260	7.1	0.1 $\underline{M}$ Na$^+$	(33)	co.;hper.54%	57
		7.8	0.05 $\underline{M}$ Na$^+$	24		57
		7.8	1.0 $\underline{M}$ Na$^+$	35		57
	275	7.8	VAR [Mg^{++}]; 0.01 $\underline{M}$ phosp.	VAR		57
	260	7.8;8.8	0.1 $\underline{M}$ Na$^+$	(26)	hper. 51%	57
	275	7.8	0.1 $\underline{M}$ Na$^+$,0.025 $\underline{M}$ phosp.	28	hper.52%;13^j	262
	280	8.8	0.5 $\underline{M}$ LiCl,2 m$\underline{M}$ gly.; 28% su.	(30)		136

Polymer	Method	pH	Medium	T_m °C	Note	Ref.

3.3.2. Copolymers

Polymer	Method	pH	Medium	T_m °C	Note	Ref.
Poly (A,br⁸A)	OD	e		68.5	A/brA = 15.6	104
	OD	e		62.5	A/brA = 4.9	104
Poly (A,ho⁸A)	OD	e		66		104
Poly (A,C)	260	5.08	0.09 $\underline{M}$ acet.		A/C/ = 1.36[f]	118
Poly (A,G)	OD	e		59.5	A/G = 2.85	104
Poly (A,br⁸G)	OD	e		71.5	A/brG = 24	104
	OD	e		61.5	A/brG = 4.56	104
	OD	e		50.5	A/brG = 3.36	104
Poly (A,ho⁸G)	OD	e		72	A/hoG = 32.3	104
Poly (A-T)	264	7.0	1 $\underline{mM}$ Na citr.	46		277
	264	7.0	VAR [Na⁺]; citr.	VAR		277
	ORD263		50% MeOH	33.5	co.;rev.	266
Poly d(A-T)	OD	6.4	0.012 $\underline{M}$ Na⁺, phosp.,EDTA	(41)	co.;d.s.hx; hper. 40%	109
	OD	7.0	1 $\underline{mM}$ Na cacd.	19.8		6
	260	7	Ag⁺, 1 $\underline{mM}$ phosp.	43	Ag/N = 0-0.8	47
		7	5 $\underline{mM}$ phosp.	40		151
	260	7	0.019 $\underline{M}$ Na⁺, 5 $\underline{mM}$ Na cacd.	51	m	194
	260	7	0.019 $\underline{M}$ Na⁺, 5 $\underline{mM}$ Na cacd.	43.2	renatured (form II)[m]	194
		7	0.1 $\underline{M}$ NaCl,0.05 $\underline{M}$ Na cacd.	66	m	193
	260	7	0.15 $\underline{M}$ Na⁺, 5 $\underline{mM}$ Na cacd.	66	m	194
	260	7	0.15 $\underline{M}$ Na⁺, 5 $\underline{mM}$ Na cacd.	60	renatured (form II)[m]	194
	260	7	0.5 $\underline{M}$ Na⁺, 5 $\underline{mM}$ Na cacd.	75.2	m	194

Polymer	Method	pH	Medium	T_m °C	Note	Ref.
Poly d(A-T) (Con't)						
	260	7	0.5 $\underline{M}$ Na$^+$, 5 m$\underline{M}$ Na cacd.	70.5	renatured (form II)^m	194
	OD	N	2 m$\underline{M}$ Na$^+$, phosp.	(24.5)		108
	260	N	1/10 SSC	44.3	hper.47.4%	143
	OD	N	1/10 SSC	45.5	co.	141
	OD	N	SSC	63	co.	141
	262	(N)	< 0.01 $\underline{M}$ Na$^+$	(40)		217
	262	(N)	0.51 $\underline{M}$ Na$^+$	(70)		217
	260	N	SSC	65.6	2.5^j	149a
	260	N	SSC; nC$_2$n	68.6	3.2^j	149a
	260	N	SSC: nC$_3$n	70.4	3.4^j	149a
	260	N	SSC; nC$_4$n	71.0	3.4^j	149a
	260	N	SSC; nC$_5$n	74.5	3.4^j	149a
	260	N	SSC; nC$_7$n	70.9	3.4^j	149a
	260	N	SSC; nC$_8$n	69.7	3.1^j	149a
	max	7.15	0.01 $\underline{M}$ KF, 1 m$\underline{M}$ Tris	49	z	149
	ORD282	7.4	0.05 $\underline{M}$ NaClO$_4$	(60-65)	1	271
	OD	7.4-7.5	VAR [Na$^+$]	VAR		109
	260	7.5	0.01 $\underline{M}$ Na citr.	(40)	co.;rev.	39
	260	7.5	VAR [Na$^+$]; citr.	VAR		39
	260	7.5	0.01 $\underline{M}$ Na citr.	39.9	2.9^n;hper.42%^{o}; co.	38
	OD		0.01 $\underline{M}$ NaCl	41.2	co.;rev.; hper.35-40%	134
	OD		0.01 $\underline{M}$ NaCl	58	co.;rev.; hper.35-40%	134
	OD		0.2 $\underline{M}$ NaCl	60	co.;rev.; hper.35-40%	334
	260	VAR	VAR [Me^{++}]	VAR		6
	OD		VAR [Na$^+$]	VAR	d.s.hx.	106
			VAR steroidal diamines	VAR	VAR N/steroid$^{\&}$	64b

Polymer	Method	pH	Medium	T_m °C	Note	Ref.
Poly d(A-s^{2}T)	OD	N	1/10 SSC	65	co.;rev.	141
	OD	N	SSC	79.5	co.;rev.	141
Poly d(A-s^{4}T)						
	260,335	N	1/10 SSC	39.3	hper.37.4(260) and 102%(335)	143
	330		2 $\underline{M}$ NaClO$_4$	54.5		125
			2 $\underline{M}$ CsSO$_4$	61.5;74	co.	142
Poly (A-U)	260	6.3	0.1 $\underline{M}$ NaCl, 5 m$\underline{M}$ Na cacd.	65.1	co.	209
	261	7.0	1 m$\underline{M}$ Na citr.	32		277
	OD	7.0	0.1 $\underline{M}$ Na$^+$, phosp.	66	d.s.hx.	289
		7	5 m$\underline{M}$ phosp.	42		272
	260	(N)	0.01 $\underline{M}$ Na citr.	49	hper.65%	52
	260	(N)	0.1 $\underline{M}$ Na citr.	62	hper. 65%	52
	260	7.5	0.01 $\underline{M}$ Na citr.	47.5	3.5^n; co.; hper.67%^{o}	38
	260	7.5	0.01 $\underline{M}$ Na citr.	(47)	co.,rev.	39
	260	7.5	VAR [Na$^+$]	VAR	co.,rev.	39
	260	8.1	0.01 $\underline{M}$ Na$_2$HPO$_4$,0.01 $\underline{M}$ Tris, 0.5 $\underline{M}$ MgCl$_2$	69		161
	OD	10.05	0.1 $\underline{M}$ Na bor.	44		289
	260		0.1 $\underline{M}$ KCl,0.01 $\underline{M}$ Na acet.	(37)	55% U;hper.40%; nco.	242
	260		0.1 $\underline{M}$ KCl,0.01 $\underline{M}$ Na acet.	(35)	69% U;hper.20%; nco.	242
	OD;ORD		0.15 $\underline{M}$ KF	(60)	co.	82
	ORD282		50% MeOH	31		266
Poly (A-fl^{5}U)	265	7.0	1 m$\underline{M}$ Na citr.	32		277
Poly (A-br^{5}U)	268	7.0	1 m$\underline{M}$ Na citr.	53		277
	OD	7.0	0.1 $\underline{M}$ Na$^+$, phosp.	83	d.s.hx.	211
	260	7.5	0.1 $\underline{M}$ Na citr.	(67)		39

Polymer	Method	pH	Medium	T_m °C	Note	Ref.
Poly (A-br^{5}U) (Con't)						
	260	7.5	VAR [Na$^+$]; citr.	VAR		39
	260	7.5	0.01 $\underline{M}$ Na citr.	67.3	2.8^n; co.; hper.64%	38
	OD	10.05	0.1 $\underline{M}$ Na bor.	24	d.s.hx.	211
	OD	VAR	0.1 $\underline{M}$ Na$^+$	VAR	d.s.hx.	211
Poly (A-dU)	260	7.5	0.01 $\underline{M}$ Na citr.	18.4	2.5^n;co.; hper.44%	38
Poly (A-Q)	262	7.0	1 $\underline{mM}$ Na citr.	58		277
Poly d(A-U)	260	7.5	0.01 $\underline{M}$ Na citr.	37.2	4.0^n;hper.44%^{o}; co.	38
	260	7.5	0.01 $\underline{M}$ Na citr.	(36)	co.;rev.	39
	260	7.5	VAR [Na$^+$]; citr.	VAR	co.; rev.	39
Poly d(A-br^{5}U)	OD	6.4	0.012 $\underline{M}$ Na$^+$,phosp.,EDTA	(50)	co.;d.s.hx.; hper.42%	109
	OD	6.4	VAR [Na$^+$]	VAR		109
	260	7.5	0.01 $\underline{M}$ Na citr.	(49)		39
	260	7.5	VAR [Na$^+$]; citr.	VAR		39
	260	7.5	0.01 $\underline{M}$ Na citr.	49.5	2.6^n;co.; hper.47%	38
	OD	N	2 $\underline{mM}$ Na$^+$, phosp.	(~34)		108
Poly (A,X)		4.0	0.05 $\underline{M}$ Na$^+$, 0.05 $\underline{M}$ acet.	82	65% A	262
		4.0	0.05 $\underline{M}$ Na$^+$,0.05 $\underline{M}$ acet.	98	87% A	262
		4.4	0.05 $\underline{M}$ Na$^+$,0.05 $\underline{M}$ acet.	66	65% A;hper.25%; 16^j	262
	255	4.4	0.05 $\underline{M}$ Na$^+$,0.05 $\underline{M}$ acet.	96	87% A; 13^j	262
	260	4.6	0.1 $\underline{M}$ Na$^+$,0.05 $\underline{M}$ acet.	70	26% A:hper.41%; 17^j	262
		4.8	0.05 $\underline{M}$ Na$^+$,0.05 $\underline{M}$ acet.	58	65% A;hper.30%; 17^j	262
	255	4.8	0.05 $\underline{M}$ Na$^+$,0.05 $\underline{M}$ acet.	79	87% A;hper.45%; 15^j	262

Polymer	Method	pH	Medium	T_m °C	Note	Ref.
Poly (A,X) (Con't)						
	255	5.2	0.05 M Na+,0.05 M acet.	61	87% A;hper.54%; 8J	262
		5.2	0.1 M Na+,0.05 M acet.	57	87% A;hper.54%; 8J	262
	260	5.6	0.1 M Na+,0.05 M acet.	41	26% A;hper.29%; 19J	262
	255	5.8	0.05 M Na+,0.05 M acet.	37	87% A;hper.27%; 8J	262
	275	7.0	0.1 M Na+,0.025 M phosp.	33	26% A;hper.27%; 8J	262
	275	7.8	0.1 M Na+,0.025 M phosp.	26	26% A;hper.45%; 15J	262
	255	7.8	0.05 M Na+,0.025 M phosp.		b.t.	262
Poly (F-T)	270	7.0	1 mM Na citr.	48		277
	270	7.0	VAR [Na+]; Na citr.	VAR		277
Poly (F-U)	266	7.0	1 mM Na citr.	33		277
	266	7.0	VAR [Na+]; Na citr.	VAR		277
Poly (F-fl^{5}U)	274	7.0	1 mM Na citr.	34		277
	274	7.0	VAR [Na+]; Na citr.	VAR		277
Poly (F-br^{5}U)	280	7.0	1 mM Na citr.	63		277
	280	7.0	VAR [Na+]; Na citr.	VAR		277
Poly (F-Q)	262	7.0	1 mM Na citr.	32		277
Poly (C,m^{4,5}C)	max	4.0	0.1 M NaCl,0.05 M acet.	(59)	83%C; ~12% hper.;b.t.	204
	max	4.0	0.1 M NaCl,0.05 M. acet.	(56)	70%C; ~13% hper.;b.t.	204
	max	4.0	0.1 M NaCl,0.05 M acet.	(55)	45%C; ~10% hper.;b.t.	204
Poly (C,ho^{4}C)	275	4.05	0.1 M NaCl,0.01 M acet.	(66)	80%C	120
	275	4.05	0.1 M NaCl,0.01 M acet.	(56)	74%C	120

Polymer	Method	pH	Medium	T_m °C	Note	Ref.
Poly ($C_{1.3}$, G)	260	7.1	0.02 M NaCl,0.2 mM EDTA, 1 mM phosp.		t. incompl.90°	118
	280	7.25	phosp.; I = 0.2 mM		t. incompl.90°	118
	280		7 mM NaCl,3.5 mM phosp.; 93% e.g.	44.8	hper.42%;17.3^j	118
Poly (C-G)	280	7.8	1 mM Na phosp.,0.1 mM EDTA	92.1	co.;rev.	122
Poly (C-I)	OD	(N)	0.1 M Na citr.	47	M = 0.35 and 0.87	48
		7.8	0.098 mM Na^+,0.5 mM EDTA, phosp.	49.5	co.	122
		7.8	VAR [Na^+],0.5 mM EDTA, phosp.	VAR	co.	122
	248	7.0	0.05 M NaCl,1 mM Na citr.	(32)	18.5% I; hper. ~50%	101
	248	7.0	0.05 M NaCl,1 mM Na citr.	(33)	30.5% I; hper. ~30%	101
	248	7.0	0.05 M NaCl,1 mM Na citr.	(28)	56.5% I; hper. ~10%	101
Poly d(C-I)	OD	7.2	0.02 M NaCl	43	d.s.hx.	69
Poly (T,U)		7	0.01 M NgCl$_2$	9.5	T/U = 8/92	248
		7	0.01 M MgCl$_2$	11.0	T/U = 14/86	248
		7	0.01 M MgCl$_2$	11.5	T/U = 22/78	248
		7	0.01 M MgCl$_2$	13.5	T/U = 37/63	248
		7	0.01 M MgCl$_2$	19.5	T/U = 71/29	248
		7	0.01 M MgCl$_2$	27.0	T/U = 86/14	248
		7	0.01 M MgCl$_2$	30.5	T/U = 93/7	248
		N			75% U; hper.76%(3.5-5.5°)	253
Poly (U,Q)	OD	7.0	0.1 M NaCl,0.01 M Mg^{++}, 0.05 M cacd.	≪20	U/Q = 2.5; no hper. >20°	73,200
Poly (U,X)	260	5.2	0.05 M Na^+,0.05 M acet.	33	20% U;hper.32%; 10^j	262

Polymer	Method	pH	Medium	T_m °C	Note	Ref.
Poly (U,X) (Con't)						
	260	5.6	0.05 $\underline{M}$ Na$^+$,0.05 $\underline{M}$ acet.	31	20% U;hper.26%; 10^j	262
	260	5.6	0.15 $\underline{M}$ Na$^+$,0.05 $\underline{M}$ acet.	35.5	20% U;hper.36%; 12^j	262
	260	6.9	0.05 $\underline{M}$ Na$^+$,0.025 $\underline{M}$ phosp., 0.05 $\underline{mM}$ sperd.	2	88% U;hper.35%; 10^j	262
	260	6.9	0.05 $\underline{M}$ Na$^+$,0.025 $\underline{M}$ phosp., 0.05 $\underline{mM}$ sperm.	18	88% U;hper.35%; 4^j	262
	275	6.9	0.1 $\underline{M}$ Na$^+$,0.025 $\underline{M}$ phosp.	18	69% U;hper.26%; 10^j	262
	275	6.9	0.1 $\underline{M}$ Na$^+$,0.025 $\underline{M}$ phosp., 0.05 $\underline{mM}$ sperm.	38	69% U;hper.22%; 9^j	262
	260	7.0	0.1 $\underline{M}$ Na$^+$,0.025 $\underline{M}$ phosp.	2.5	84% U;hper.40%; 10^j	262
	260	7.0	0.1 $\underline{M}$ Na$^+$,0.025 $\underline{M}$ phosp., 0.05 $\underline{mM}$ sperm.	19	84% U;hper.40%; 14^j	262
	275	7.0	0.1 $\underline{M}$ Na$^+$,0.025 $\underline{M}$ phosp.	31	20% U;hper.50%; 13^j	262
	275	7.8	0.1 $\underline{M}$ Na$^+$,0.025 $\underline{M}$ phosp.	31	20% U;hper.57%; 17^j	262

Polymer	Method	pH	Medium	T_m °C	Note	Ref.
		3.3.3.	Sugar-Phosphate Backbone Analogues			
Poly (A$_\mathrm{s}$U)	260	8.1	0.01 $\underline{M}$ Na$_2$HPO$_4$,0.5 $\underline{M}$ MgCl$_2$,0.01 $\underline{M}$ Tris	68		161
Poly ($_\mathrm{s}$A$_\mathrm{s}$U)		7.5	0.01 $\underline{M}$ Na citr.	49		165
	260		0.01 $\underline{M}$ Na$^+$, citr.	49	hper.65%	52
	260		0.1 $\underline{M}$ Na$^+$, citr.	62	hper.65%	52
Poly (C$_\mathrm{s}$I)	OD	(N)	0.1 $\underline{M}$ Na citr.	50		48
Poly ($_\mathrm{s}$C-I)		7.5	0.1 $\underline{M}$ Na citr.	57		165
Poly ($_\mathrm{s}$I$_\mathrm{s}$C)	OD	(N)	0.1 $\underline{M}$ Na citr.	51		48

Polymer	Method	pH	Medium	T_m °C	Note	Ref.

3.3.4. Complexes

Poly A Complexes

Polymer	Method	pH	Medium	T_m °C	Note	Ref.
2 Poly (A,X)	255	7.0	0.2 $\underline{M}$ Na$^+$,0.025 $\underline{M}$ phosp. (51)		A/X = 26/74	262
Poly C$^+$		5	VAR [Na$^+$]	VAR		76
Poly ho^{4}C	OD	7.2	0.5 $\underline{M}$ NaCl,0.05 $\underline{M}$ Na cacd., 0.1 $\underline{M}$ MgCl$_2$	32		234
Poly mh^{4}C	OD	7.2	0.5 $\underline{M}$ NaCl,0.05 $\underline{M}$ Na cacd., 0.1 $\underline{M}$ MgCl$_2$	35		234
Poly (C,U)	260	7.0	0.1 $\underline{M}$ NaCl,0.01 $\underline{M}$ phosp.	50	U/C = 84/16	120
Poly (ho^{4}C,U)	260	7.0	0.1 $\underline{M}$ NaCl,0.01 $\underline{M}$ phosp.	43	U/hoC = 76/24	120
Poly I	260	6.0	0.15 $\underline{M}$ Na$^+$,0.01 $\underline{M}$ Mg^{++}	52.0		49
	260	7.0	0.15 $\underline{M}$ Na$^+$,0.01 $\underline{M}$ Mg^{++}	48.0		49
	248,260	7.0	0.15 $\underline{M}$ Na$^+$	42.5		49
	OD	7.0	0.15 $\underline{M}$ Na$^+$	45.2		180
	250	7	0.1 $\underline{M}$ NaCl,0.05 $\underline{M}$ Na cacd.	43.5		176
	255	7	0.1 $\underline{M}$ NaCl,0.05 $\underline{M}$ Na cacd., 0.01 $\underline{M}$ MgCl$_2$	53.7		176
	OD	7.0	0.15 $\underline{M}$ Na$^+$	44.8		178
	OD	7.05	0.02 $\underline{M}$ Na$^+$, Na phosp.	30.2		178
2 Poly I	max	6	1.0 $\underline{M}$ NaCl,2 m$\underline{M}$ phosp. (55)		b.t.	50
	254	6.15	0.1 $\underline{M}$ Na$^+$,phosp. 40.3$\pm$0.2		hper.30%[a]	64
	OD	6.9	0.1 $\underline{M}$ NaCl,0.01 $\underline{M}$ Na cacd.	39	[a]	228
	254; ORD294,252	7.4	0.1 $\underline{M}$ NaCl,0.01 $\underline{M}$ glycylgly (40)		rev.	214
Poly T	266.5	6.9	SSC	79	rev.	255
	OD	7.0	0.1 $\underline{M}$ NaCl,0.05 $\underline{M}$ Na cacd. (78)			170

Polymer	Method	pH	Medium	T_m °C	Note	Ref.

Poly A Complexes (Con't)

  Poly T (Con't)

Polymer	Method	pH	Medium	T_m °C	Note	Ref.
	260	7	SSC	71	hper.38.6%	218
		7	SSC	79		256
		7	SSC	79.0		248
	265	N	0.2 $\underline{M}$ Na$^+$	75	hper.48%	265
	265	N	0.1 $\underline{M}$ Na$^+$	66	hper.52%	265
	265	N	0.01 $\underline{M}$ Na$^+$	51	hper.50%	265
	265	N	< 0.01 $\underline{M}$ Na$^+$	29	hper.15%	265
		7.5	VAR [Na$^+$];0.01 $\underline{M}$ phosp.	VAR		257
	260	7.8	0.082 $\underline{M}$ Na$^+$,0.01 $\underline{M}$ phosp., VAR [Mg^{++}],VAR concn. of sperm. or sperd.	VAR		251
	262.5		VAR [Na$^+$], [Mg^{++}], and [nC$_z$n]	VAR	z ≤ 6	251
	260; ORD289		30% MeOH	−7.5		266
	260; ORD289		40% MeOH	−9		266
	260; ORD289		50% MeOH	−12		266
	260; ORD289		60% MeOH	−1		266
	260; ORD289		70% MeOH	−12.5	hysteresis	266
	260; ORD289		80% MeOH	−17		266
2 Poly T	260	7	0.01 $\underline{M}$ Na$^+$	59.7	hper.53%	174
	260	7	0.1 $\underline{M}$ Na$^+$	78.2	hper.65.0%	174
	260	7	0.15 $\underline{M}$ Na$^+$	80.5	hper.73.7%	174
	265	7.2	SSC	80	hper.70%	265
	260	7.2;7.4	0.1 $\underline{M}$ Na$^+$,0.01 $\underline{M}$ phosp.	72		265
	265	7.2	0.15 $\underline{M}$ NaCl,0.01 $\underline{M}$ phosp.	80	hper.70%	265

Polymer	Method	pH	Medium	T_m °C	Note	Ref.
Poly A Complexes (Con't)						
2 Poly T (Con't)						
	265	7.2	0.01 $\underline{M}$ NaCl	56	hper.43%	265
Poly dT	OD	6.30	0.025 $\underline{M}$ Na phosp.	52.9$\pm$0.1		64a
	OD	6.30	0.025 $\underline{M}$ Na phosp., 2 m$\underline{M}$ n$\bar{C}_z$n	VAR	z = 2-7^w	64a
	OD	7.0	0.05 $\underline{M}$ NaClO$_4$	59	hper.40-50%	271
	IMMUN	7.4	0.14 $\underline{M}$ NaCl,0.01 $\underline{M}$ Tris	(76)		244
	OD	7.8	0.1 $\underline{M}$ phosp.	68.5		37
			VAR steroidal diamines	VAR	VAR N/steroid$^\&$	64b
Poly (T,U)		7	SSC	60.5	T/U = 8/92	248
		7	SSC	61	T/U = 14/86	248
		7	SSC	62	T/U = 22/78	248
		7	SSC	63.0	T/U = 37/63	248
		7	SSC	66.0	T/U = 71/29	248
		7	SSC	71.0	T/U = 86/14	248
		7	SSC	75.0	T/U = 93/7	248
Poly U	260	4.60-5.05			d	45
	max	5.5	0.05 $\underline{M}$ KF,1 m$\underline{M}$ EDTA	44.2		70
	max	6	1.0 $\underline{M}$ NaCl,2 m$\underline{M}$ phosp.	(80)	b.t.	50
	257,280	6.15	0.1 $\underline{M}$ Na$^+$, phosp.	56.0$\pm$0.2	hper.51%^{a}	64
	OD	6.30	0.025 $\underline{M}$ Na phosp.	46.6$\pm$0.2		64a
	OD	6.30	0.025 $\underline{M}$ Na phosp., 0.1 m$\underline{M}$, n$\bar{C}_z$n	VAR	z = 2-4^w	64a
		6.5	0.1 $\underline{M}$ KCl,0.01 $\underline{M}$ Na acet.	VAR	VAR M$_A$	242
	254	6.5	0.068 $\underline{M}$ NaCl,1 m$\underline{M}$ Na acet.	(35)		242
	254	6.5	0.095 $\underline{M}$ NaCl,1 m$\underline{M}$ Na acet.	(39)		242
	254	6.5	0.205 $\underline{M}$ NaCl,1 m$\underline{M}$ Na acet.	(46)		242
	254	6.5	0.308 $\underline{M}$ NaCl,1 m$\underline{M}$ Na acet.	(50)		242

Polymer	Method	pH	Medium	T_m °C	Note	Ref.
Poly A Complexes (Con't)						
Poly U (Con't)						
	254	6.5	0.545 $\underline{M}$ NaCl, 1 m$\underline{M}$ Na acet. (54)			242
	259	6.5-7.0	0.05 $\underline{M}$ NaCl,1 m$\underline{M}$ Ma citr. (50)			88
	259	6.5-7.0	0.05 $\underline{M}$ NaCl,1 m$\underline{M}$ Na citr., (81) 0.01 $\underline{M}$ nC$_2$n			88
	259	6.5-7.0	0.05 $\underline{M}$ NaCl,1 m$\underline{M}$ Na citr., (80) 0.01 $\underline{M}$ nC$_3$n			88
	259	6.5-7.0	0.05 $\underline{M}$ NaCl,1 m$\underline{M}$ Na citr., (76) 0.01 $\underline{M}$ nC$_4$n			88
	259	6.5-7.0	0.05 $\underline{M}$ NaCl,1 m$\underline{M}$ Na citr., 0.01 $\underline{M}$ nC$_5$n (72)			88
	259	6.5-7.0	0.05 $\underline{M}$ NaCl,1 m$\underline{M}$ Na citr., (67) 0.01 $\underline{M}$ n C$_6$n			88
	259	6.5-7.0	0.05 $\underline{M}$ NaCl,1 m$\underline{M}$ Na citr., (53) 0.01 $\underline{M}$ nC$_8$n			88
	CAL	6.8	0.50 $\underline{M}$ Na$^+$,0.01 $\underline{M}$ citr.	54.3;72.1	2→3; 3→1	1
	266.5	6.9	SSC	59	rev.	255
	OD;CAL	6.8-6.9	5-10 m$\underline{M}$ cacd.,VAR [Na$^+$]	VAR		130
	OD;CAL	6.8-6.9	5-10 m$\underline{M}$ cacd.,VAR [K$^+$]	VAR		130
	OD	6.8-7.0	VAR [Na$^+$],[K$^+$], and [Mg^{++}]; 0.01 $\underline{M}$ cacd.	VAR	2→1; 2→3, VAR[A]/[U]	129
	260	6.95	0.1 $\underline{M}$ NaCl,0.01 $\underline{M}$ cacd.	56.1	no hper. at 280 mμ	243
	260	6.95	0.1 $\underline{M}$ NaCl,0.01 $\underline{M}$ cacd., 1 m$\underline{M}$ Mg^{++}	67.2		243
	280	6.95	0.1 $\underline{M}$ NaCl,0.01 $\underline{M}$ cacd., 1 m$\underline{M}$ MgCl$_2$	67.2	OD decreases at 49.5°	243
	OD	7.0	0.05 $\underline{M}$ NaCl$_4$	51	hper.40-50%	271
	OD	7.0	0.1 $\underline{M}$ NaCl,0.05 $\underline{M}$ Na cacd. (60)			170
	260	7.0	0.1 $\underline{M}$ NaCl,0.01 $\underline{M}$ phosp.	56		120
	260	7.0	0.1 $\underline{M}$ NaCl,0.01 $\underline{M}$ phosp.	56		120
	OD	7.0	0.1 $\underline{M}$ Na$^+$, phosp.	57		211
	260	7.0	0.1 $\underline{M}$ NaCl,0.05 $\underline{M}$ Na cacd., 0.01 $\underline{M}$ MgCl$_2$	76.7		177

Polymer	Method	pH	Medium	T_m °C	Note	Ref.
Poly A Complexes (Con't)						
Poly U (Con't)						
	OD	7.0	0.1 $\underline{M}$ Na$^+$,0.025 $\underline{M}$ phosp.	54	2 → 1	262
	OD	7.0	0.1 $\underline{M}$ Na$^+$,0.025 $\underline{M}$ phosp., 1 m$\underline{M}$ sperm.	39	2 → 3	262
	260	7.0	0.15 $\underline{M}$ Na$^+$	57.5		49
	OD	7.0	1 $\underline{M}$ Na$^+$,0.025 $\underline{M}$ phosp.	45	2 → 3	262
		7	4 m$\underline{M}$ NaCl,1 m$\underline{M}$ Na citr.	37		151
	259	7	4 m$\underline{M}$ NaCl,1 m$\underline{M}$ Na citr.	35.5	2 → 1	89
	259	7	4 m$\underline{M}$ NaCl,1 m$\underline{M}$ Na citr., poly-L-lysine	36,81	2 → 1; 3 → 1; Lys/N = 0.25 to 0.66	89
	259	7	0.04 $\underline{M}$ NaCl,0.01 $\underline{M}$ cacd.	51	M_U = 0.5; 2→1	12
	259	7	0.04 $\underline{M}$ NaCl,0.01 $\underline{M}$ cacd.	51	M_U = 0.67; 2→1	12
	260	7	0.1 $\underline{M}$ NaCl,0.01 $\underline{M}$ MgCl$_2$, 0.05 $\underline{M}$ cacd.	69	hper. 51%	203
	259	7	0.1 $\underline{M}$ NaCl,1 m$\underline{M}$ Na citr.	(54)		88
	259	7	0.1 $\underline{M}$ NaCl,1 m$\underline{M}$ Na citr., 1 m$\underline{M}$ sperm.	83		88
	280	7	0.1 $\underline{M}$ NaCl,1 m$\underline{M}$ Na citr., 1 m$\underline{M}$ sperm.	39;83	2→3; 3→1	88
	OD	7	VAR [Na$^+$]	VAR		88
		7	1/10 SSC	43		256
		7	SSC	59	hper.53%	256
		7	SSC	60		248
	260	7	SSC	57	hper.24%	218
	OD	7	SSC	(58)	co.	50
	ORD	7	SSC	(60)	co.	50
	259	7	0.2 $\underline{M}$ NaCl,0.01 $\underline{M}$ cacd.	65	M_U = 0.5;2→1	12
	260	7	VAR [Na$^+$]		T_m 2-1 = 348.7+ 19.6 log[Na$^+$]	152
		7	KCl,citr.,cacd.;I = 0.05 $\underline{M}$	46.5		279

Polymer	Method	pH	Medium	T_m °C	Note	Ref.
Poly A Complexes (Con't)						
Poly U (Con't)						
		7	KCl,citr.,cacd.; I = 0.2 $\underline{M}$	65.0		279
		7	KCl,citr.,cacd.; I = 0.6 $\underline{M}$	75.5		279
		7	KCl,citr.,cacd.; I = 1.0 $\underline{M}$	79.5		279
		7	KCl,citr.,cacd.; I = 2.0 $\underline{M}$	88.6		279
		7	KCl,citr.,cacd.; I = 2.4 $\underline{M}$	92.0		279
	COND	N	0.05 $\underline{M}$ KCl	(44)	$M_A \sim M_U \sim 0.5$	100
	COND	N	0.05 $\underline{M}$ KCl	(49.5)	$M_A \sim 3$; $M_U \sim 1$	100
280,283.5 287	N		0.50 $\underline{M}$ Na$^+$	54	$2 \rightarrow 3$; $2 \rightarrow 1$	10
259,280, 287	N		0.50 $\underline{M}$ Na$^+$	72.5	$3 \rightarrow 1$	10
	N		VAR [Na$^+$]	VAR	$2 \rightarrow 3$	10
EB	N		Mg^{++}	(64)	Mg/(A+U) = 1	117
259	(N)		I = 0.05 $\underline{M}$	47.5		202
259	(N)		histone; I = 0.05 $\underline{M}$	48.5	H/(A+U) = 1	202
	N		SSC	58.5		253
max		7.15	0.01 $\underline{M}$ KF,1 m$\underline{M}$ Tris	37.4	z	149
259		7.2	0.1 $\underline{M}$ NaCl,0.05 $\underline{M}$ Na cacd.	58.7		181
259		7.2	0.1 $\underline{M}$ NaCl,0.05 $\underline{M}$ Na cacd.; D$_2$0	59.3		181
260		7.2	0.15 $\underline{M}$ NaCl,0.01 $\underline{M}$ MgCl$_2$, 0.015 Tris	76.5		125,126
		7.25	0.1 $\underline{M}$ Tris	57.0±0.2	hper.36%^{a}	63
OD		7.4	5 m$\underline{M}$ NaCl,0.01 $\underline{M}$ Tris	(47)		15
IMMUN		7.4	0.14 $\underline{M}$ NaCl,0.01 $\underline{M}$ Tris	(55)		244
OD		7.4	VAR [Na$^+$]; phosp.	VAR	$3 \rightarrow 1$; $2 \rightarrow 1$	102
259		7.5	cacd.	58	co.	215
ORD286		7.5	0.15 $\underline{M}$ KF	(58)	b.t.	215
255		7.8	0.05 $\underline{M}$ Na$^+$, 0.025 $\underline{M}$ phosp.	(50)	$2 \rightarrow 1$	262
max		7.8	0.1 $\underline{M}$ Na$^+$,0.01 $\underline{M}$ phosp.	57	hper.48%	289
OD		7.8	0.1 $\underline{M}$ phosp.	56.8		37

Polymer	Method	pH	Medium	T_m °C	Note	Ref.

Poly A Complexes (Con't)

Poly U (Con't)

Polymer	Method	pH	Medium	T_m °C	Note	Ref.
	260	7.8	0.082 $\underline{M}$ Na$^+$,0.01 $\underline{M}$ phosp., VAR[Mg^{++}],[sperm.],and [sperd.]	VAR		251
	262.5	7.8	0.01 $\underline{M}$ phosp.,VAR[Na$^+$], [Mg^{++}],and [nC$_2$n]	VAR		251
	260	8.1	0.2 $\underline{M}$ Na$^+$,0.04 $\underline{M}$ pyrophosp.	61		96
	OD	9.95	0.03 $\underline{M}$ Na$^+$	36.1		152
	OD	9.95	0.07 $\underline{M}$ Na$^+$	36.8		152
	OD	9.95	0.2 $\underline{M}$ Na$^+$	35.6		152
	OD	9.95	0.4 $\underline{M}$ Na$^+$	33.4		152
	OD	9.95	1.0 $\underline{M}$ Na$^+$	28.1		152
	OD	10.05	0.1 $\underline{M}$ Na bor.	35		211
	CD		0.15 $\underline{M}$ KF	61	co.	82
	1625,1657		0.14 $\underline{M}$ Na$^+$;D$_2$0	(58)		183
	1657		0.16 $\underline{M}$ Na$^+$,0.02 $\underline{M}$ Mg^{++}; D$_2$0	(44;67)		183
	1657		0.44 $\underline{M}$ Na$^+$;D$_2$0	(54;68)		183
	260	VAR		VAR		61
	OD	VAR	VAR[Na$^+$]	VAR		153
	max		e.g.	14.8		70
	ORD282		50% MeOH	-25		266
			VAR steroidal diamines	VAR	VAR N/steroid$^{\&}$	64b
2 Poly U	OD	6.30	0.025 $\underline{M}$ Na phosp., 2 m$\underline{M}$ nC$_2$n	VAR	z = 2-4^w	64a
	OD	6.4	0.12 $\underline{M}$ NaCl,acet.	57.7	3 → 1	152
	OD	6.8-7.0	VAR[Na$^+$] and [K$^+$]; 0.01 $\underline{M}$ cacd.	VAR	3→2; 3→1 VAR[A]/[U]	125
	260	6.9	0.05 $\underline{M}$ Na$^+$,0.05 $\underline{M}$ Na cacd.	49.8	co.	152
	260	6.9	0.075 $\underline{M}$ Na$^+$,0.05 $\underline{M}$ Na cacd.	53.5	co.	152

Polymer	Method	pH	Medium	T_m $^\circ C$	Note	Ref.

Poly A Complexes (Con't)

  2 Poly U (Con't)

	260	6.9	0.1 $\underline{M}$ Na$^+$,0.05 $\underline{M}$ Na cacd.	55.2	co.	152
	260	6.95	0.05 $\underline{M}$ NaCl,0.01 $\underline{M}$ cacd.	45.0;53.5		243
	280	6.95	0.05 $\underline{M}$ NaCl,0.01 $\underline{M}$ cacd.	45.2		243
	260	6.95	VAR[Na$^+$]	VAR	2→1;3→1;2→3	243
	260	6.95	0.1 $\underline{M}$ NaCl,0.01 $\underline{M}$ cacd., 1 m$\underline{M}$ MgCl$_2$	67.7		243
	280	6.95	0.1 $\underline{M}$ NaCl,0.01 $\underline{M}$ cacd., 1 m$\underline{M}$ MgCl$_2$	67.0		243
	259; ORD400	7.0	I = 0.0155 $\underline{M}$	(23);(41)	3→2; 2→1	60
	259	7.0	0.1 $\underline{M}$ NaCl,0.1 $\underline{M}$ Tris	(56)	$10^{-5}\underline{M}$ N;co.	192
	ORD350	7.0	0.1 $\underline{M}$ NaCl,0.1 $\underline{M}$ Tris	(59)	$10^{-2}\underline{M}$ N;co.	192
	260	7.0	0.1 $\underline{M}$ NaCl,0.01 $\underline{M}$ Na cacd.	(56)		104
	OD	7.0	0.1 $\underline{M}$ NaCl,0.05 $\underline{M}$ cacd.	60		200
	OD	7.0	0.1 $\underline{M}$ NaCl,0.05 $\underline{M}$ cacd., 0.01 $\underline{M}$ MgCl$_2$	76		200
	260	7.0	0.1 $\underline{M}$ NaCl,0.01 $\underline{M}$ Na cacd., 0.01 $\underline{M}$ MgCl$_2$	(72)		104
	OD	7.0	0.1 $\underline{M}$ NaCl,0.05 $\underline{M}$ Na cacd.	59		177
	OD	7.0	0.1 $\underline{M}$ Na$^+$,0.025 $\underline{M}$ phosp.	51	3 → 2	262
	OD	7.0	0.1 $\underline{M}$ Na$^+$,0.025 $\underline{M}$ phosp., 1 m$\underline{M}$ sperm.	83	3 → 1	262
	OD	7.0	1 $\underline{M}$ Na$^+$,0.025 $\underline{M}$ phosp.	85	3 1	262
	OD	7.0	0.15 $\underline{M}$ Na$^+$	60		178
		7	0.01 $\underline{M}$ NaCl,1 m$\underline{M}$ Na citr.	25;38		272
	259	7	0.01 $\underline{M}$ NaCl,1 m$\underline{M}$ Na citr.	39		88
	259	7	0.01 $\underline{M}$ NaCl,1 m$\underline{M}$ Na citr.	26;39	after melting and slow cooling	88
	280	7	0.01 $\underline{M}$ NaCl,1 m$\underline{M}$ Na citr.	---		88
	280	7	0.01 $\underline{M}$ NaCl,1 m$\underline{M}$ Na citr.	26	after melting and slow cooling	88

Polymer	Method	pH	Medium	T_m $°C$	Note	Ref.

Poly A Complexes (Con't)

2 Poly U (Con't)

Polymer	Method	pH	Medium	T_m $°C$	Note	Ref.
	259	7	0.01 $\underline{M}$ NaCl,1 m$\underline{M}$ Na citr.	(24);(38)		88
	259	7	0.01 $\underline{M}$ NaCl,1 m$\underline{M}$ Na citr., 1 m$\underline{M}$ nC$_2$n	72		88
	259	7	0.01 $\underline{M}$ NaCl,1 m$\underline{M}$ Na citr.	26;37.5	3 → 2; 2 → 1	89
	280	7	0.01 $\underline{M}$ NaCl,1 m$\underline{M}$ Na citr.	26		89
	259,280	7	0.01 $\underline{M}$ NaCl,1 m$\underline{M}$ Na citr., poly-L-lysine	81.5	3 → 1	89
	259	7	0.01 $\underline{M}$ NaCl,1 m$\underline{M}$ Na citr., 1 m$\underline{M}$ nC$_3$n	(72)		88
	259	7	0.01 $\underline{M}$ NaCl,1 m$\underline{M}$ Na citr., 1 m$\underline{M}$ nC$_4$n	(69)		88
	259	7	0.01 $\underline{M}$ NaCl,1 m$\underline{M}$ Na citr., 1 m$\underline{M}$ nC$_5$n	(58)		88
	259	7	0.01 $\underline{M}$ NaCl,1 m$\underline{M}$ Na citr., 1 m$\underline{M}$ nC$_6$n	(52)		88
	259	7	SSC	(60)	3 → 1	10,60
	259	7	0.04 $\underline{M}$ NaCl,0.01 $\underline{M}$ cacd.	40.5	M_U = 0.67;3→2	12
	259	7	0.2 $\underline{M}$ NaCl,0.01 $\underline{M}$ cacd.	65	M_U = 0.67;3→2	12
	259	7	0.5 $\underline{M}$ NaCl,0.01 $\underline{M}$ cacd.	72.5	M_U = 0.5;3→2	12
	OD	7	0.01 $\underline{M}$ Na$^+$,Na cacd.	(21;38)		154
	260	7	0.025 $\underline{M}$ Na$^+$, Na cacd.	(33;47)		154
	260	7	0.095 $\underline{M}$ Na$^+$, Na cacd.	(55;58)		154
	260	7	0.495 $\underline{M}$ Na$^+$, Na cacd.	(75)		154
	260	7	0.995 $\underline{M}$ Na$^+$, Na cacd.	(82)		154
	260	7	1 $\underline{M}$ LiCl,0.05 $\underline{M}$ Na cacd.; 28% su.	74.6	b.t.	154
	260	7	1 $\underline{M}$ NaCl,0.05 $\underline{M}$ Na cacd.; 28% su.	73.7	b.t.	154
	260	7	1 $\underline{M}$ KCl,0.05 $\underline{M}$ Na cacd.; 28% su.	72.3	b.t.	154
	260	7	1 $\underline{M}$ RbCl,0.05 $\underline{M}$ Na cacd.; 28% su.	71.5	b.t.	154

Polymer	Method	pH	Medium	T_m °C	Note	Ref.

Poly A Complexes (Con't)

2 Poly U (Con't)

Polymer	Method	pH	Medium	T_m °C	Note	Ref.
	260	7	1 M CsCl,0.05 M Na cacd.; 28% su.	71.5	b.t.	154
	260	7	VAR[Na$^+$]		$T_m(3\text{-}1) = 353.1 + 25.6 \log[\text{Na}^+]$	152
	260	7	VAR[Na$^+$]		$T_m(2\text{-}3) = 322.1 - 15.3 \log[\text{Na}^+]$	152
		7	VAR[Na$^+$]	VAR	$T_m(2\text{-}1) = 20.0 \cdot \log[\text{Na}^+] + 76.2$	10,60
		7	VAR[Na$^+$]	VAR	$T_m(3\text{-}1) = 26.1 \cdot \log[\text{Na}^+] + 80.3$	10,60
		7	VAR[Na$^+$]	VAR	$T_m(3\text{-}2) = 30.8 \cdot \log[\text{Na}^+] + 81.6$	10,60
	OD	7	VAR[Na$^+$]	VAR		88
		7	1 mM MgCl$_2$,1 mM Na citr.	70		272
	259,280	7	1 mM MgCl$_2$,1 mM Na citr.	70	3 → 1	22
	259,280	7	1 mM MgCl$_2$,1 mM Na citr., poly-L-lysine	86		22
	260	(7)	0.084 M NaCl,0.066 M K phosp.	55	no hper. at 280 mμ	219
	280,283.5	N	0.040 M Na$^+$	38.4	3 → 2	10
	283.5,287	N	0.040 M Na$^+$	48.5	2 → 1	10
	OD	7.1	0.02 M Na$^+$	42.7;46.7	traces of Mg^{++}	178
	259	7.5±0.2	6.5 mM Na$^+$	32.2	2 → 1; U/A = 2	9
	259	7.5±0.2	9.6 mM Na$^+$	36.0	2 → 1; U/A = 2	9
	259	7.5±0.2	10.5 mM Na$^+$	36.5	2 → 1; U/A = 2	9
	259	7.5±0.2	11 mM Na$^+$	21.9;37.6	3 → 2; 2→1; U/A = 2	9
	259; ORD284	7.5	0.15 M KF	(59)	co.;rev.	215
	255	7.8	0.3 M Na$^+$,0.025 M phosp.	70		262

Polymer	Method	pH	Medium	T_m °C	Note	Ref.

Poly A Complexes (Con't)

 2 Poly U (Con't)

Polymer	Method	pH	Medium	T_m °C	Note	Ref.
	262.5	7.8	VAR[Na$^+$],[Mg^{++}], and [nC$_2$n]; VAR 0.01 $\underline{M}$ phosp.		$z \leq 6$	251
	OD	8.15	0.12 $\underline{M}$ NaCl,bor.	56.9	3 → 1	152
	OD	8.6	0.12 $\underline{M}$ NaCl,bor.	54.8	3 → 1	152
	OD	8.9	0.12 $\underline{M}$ NaCl,bor.	53.4	3 → 1	152
	OD	9.15	0.12 $\underline{M}$ NaCl,bor.	51.0	3 → 1	152
	OD	9.4	0.12 $\underline{M}$ NaCl,bor.	49.1	3 → 1	152
	OD	9.6	0.12 $\underline{M}$ NaCl,bor.	45.6	3 → 1	152
	OD	9.8	0.12 $\underline{M}$ NaCl,bor.	40.6	3 → 1	152
	OD	10.0	0.12 $\underline{M}$ NaCl,bor.	35.0	3 → 1	152
	OD	10.1	0.12 $\underline{M}$ NaCl,bor.	33.5	3 → 1	152
	OD	10.2	0.12 $\underline{M}$ NaCl,bor.	30.0	3 → 1	152
	260	10	VAR[Na$^+$];bor.	VAR		152
	OD	9.95	0.03 $\underline{M}$ Na$^+$	18.7	3 → 2	152
	OD	9.95	0.07 $\underline{M}$ Na$^+$	21.0	3 → 2	152
	OD	9.95	0.2 $\underline{M}$ Na$^+$	21.0	3 → 2	152
	OD	9.95	0.4 $\underline{M}$ Na$^+$	19.4	3 → 2	152
	OD	9.95	1.0 $\underline{M}$ Na$^+$	16.5	3 → 2	152
	260	VAR	0.15 and 0.5 $\underline{M}$ NaCl	VAR		152
	OD	VAR	VAR[Na$^+$]	VAR		153
	IR		VAR[Na$^+$]	VAR		182
	IR		VAR[Na$^+$]	VAR		98
	ORD282		50% MeOH	−27		266
Poly (U,m^{3}U)	N		SSC	34	hper.17.2%; U/mU = 3	253
Poly (U,hm^{5}U)	260	7	SSC	33	hper.17%; U/hmU = 4	218

Polymer	Method	pH	Medium	T_m °C	Note	Ref.
Poly A Complexes (Con't)						
Poly (U,hm^{5}U) (Con't)						
	260	7	SSC	53	hper.16%; U/hmU = 2	218
Poly (U,s^{4}U)						
	260,330	6.9	VAR[Na$^+$];0.05 $\underline{M}$ Na cacd.	VAR	U/sU = 1.25 to 11	221
2 Poly (U,s^{4}U)						
	260	(7)	0.084 $\underline{M}$ NaCl,0.066 $\underline{M}$ K phosp.	34.5	U/sU = 6	219
	330	(7)	0.084 $\underline{M}$ NaCl,0.066 $\underline{M}$ K phosp.	20;37.5	U/sU = 6	219
	260	(7)	0.084 $\underline{M}$ NaCl,0.066 $\underline{M}$ K phosp.	39	U/sU = 13	219
	330	(7)	0.084 $\underline{M}$ NaCl,0.066 $\underline{M}$ K phosp.	28;42	U/sU = 13	219
Poly (U,h^{5,6}U)						
	1658,1624		0.15 $\underline{M}$ Na$^+$; D$_2$0	50	U/hU = 19; pD = 7	36
	1658,1624		0.15 $\underline{M}$ Na$^+$; D$_2$0	47	U/hU = 6.7; pD = 7	36
	1658,1624		0.15 $\underline{M}$ Na$^+$; D$_2$0	23	U/hU = 2.6; pD = 7	36
Poly (U,X)	255	7.0	0.1 $\underline{M}$ Na$^+$,0.025 $\underline{M}$ phosp.	42	U/X = 5.25; 2 → 1	262
	255	7.0	0.1 $\underline{M}$ Na$^+$,0.025 $\underline{M}$ phosp.	45	U/X = 7.35; 2 → 1	262
	255	7.0	1 $\underline{M}$ Na$^+$,0.025 $\underline{M}$ phosp.	50	U/X = 5.25; 2 → 3	262
2 Poly (U,X)	260	7.0	0.1 $\underline{M}$ Na$^+$,0.025 $\underline{M}$ phosp.	(32)	U/X = 2.23	262
	255,280	7.0	0.1 $\underline{M}$ Na$^+$,0.025 $\underline{M}$ phosp.	32	U/X = 5.25 3 → 2	262
	255	7.0	0.1 $\underline{M}$ Na$^+$,0.025 $\underline{M}$ phosp.	30	U/X = 7.35 3 → 2	262
	255	7.0	1 $\underline{M}$ Na$^+$,0.025 $\underline{M}$ phosp.	58	U/X = 5.25 3 → 1	262

Polymer	Method	pH	Medium	T_m °C	Note	Ref.
Poly A Complexes (Con't)						
2 Poly (U,X) (Con't)						
255		7.0	0.1 $\underline{M}$ Na$^+$, 0.025 $\underline{M}$ phosp., 0.05 m$\underline{M}$ sperm.	38;57	U/X = 5.25; 2 3;3 1	262
255		7.0	0.1 $\underline{M}$ Na$^+$,0.025 $\underline{M}$ phosp., 0.05 m$\underline{M}$ sperm.	46;71	U/X = 7.35; 2 3;3 1	262
Poly m^{3}U	(N)				no hper. >3°	253
Poly e^{5}U			0.15 $\underline{M}$ NaCl		t.s.hx.	226
2 Poly e^{5}U	260	N	0.01 $\underline{M}$ Na$^+$	47	co.	247
	260	N	0.02 $\underline{M}$ Na$^+$	(49)	co.	247
	260	N	0.05 $\underline{M}$ Na$^+$	(55)	co.	247
	260	N	0.1 $\underline{M}$ Na$^+$	60	co.	247
	260	N	0.5 $\underline{M}$ Na$^+$	(74)	co.	247
	260	N	1.0 $\underline{M}$ Na$^+$	(80)	co.	247
	260	N	VAR[Na$^+$]	VAR		247
Poly hm^{5}U	260	7	SSC	30.5	hper.16%	218
Poly fl^{5}U	263	6-7	SSC	55	hper.50%	256
		7	1/10 SSC	40		256
	OD	VAR	0.15 $\underline{M}$ Na$^+$	VAR	strandedness unspecified	170
	OD	VAR	0.15 $\underline{M}$ Na$^+$,0.01 $\underline{M}$ Mg^{++}	VAR		170
2 Poly fl^{5}U		7.0	0.1 $\underline{M}$ NaCl,0.05 $\underline{M}$ Na cacd.	49.6		177
2 Poly cl^{5}U						
269		6.5			hper.35.4%	171
		7.0	0.1 $\underline{M}$ NaCl,0.05 $\underline{M}$ Na cacd.	86.1		177
	OD	7.0	0.1 $\underline{M}$ NaCl,0.05 $\underline{M}$ Na cacd.	(86)	strandedness unspecified	170
		7	SSC	82		256

Polymer	Method	pH	Medium	T_m $^\circ C$	Note	Ref.
Poly A Complexes (Con't)						
Poly br^{5}U		7	1/10 SSC	69		256
	OD	10.0	0.1 $\underline{M}$ Na bor.	13		211
	OD	VAR	0.1 $\underline{M}$ Na$^+$	VAR		211
2 Poly br^{5}U						
	269	6.5			hper.38.5%	171
	OD	7.0	0.1 $\underline{M}$ Na$^+$	91		211
	OD	7.0	0.1 $\underline{M}$ NaCl,0.05 $\underline{M}$ Na cacd.	(95)	strandedness unspecified	170
		7.0	0.1 $\underline{M}$ NaCl,0.05 $\underline{M}$ Na cacd.	95.1		177
		7	SSC	87		256
	260,280	7.5	0.1 $\underline{M}$ Na$^+$,0.01 $\underline{M}$ phosp.	87.5		211
	OD	VAR	0.1 $\underline{M}$ Na$^+$	VAR		211
2 Poly io^{5}U		7.0	0.1 $\underline{M}$ NaCl,0.05 $\underline{M}$ Na cacd.	94.8		177
	OD	7.0	0.1 $\underline{M}$ NaCl,0.05 $\underline{M}$ Na cacd.	(94)	strandedness unspecified	170
Poly ho^{5}U	OD	7.0	0.1 $\underline{M}$ NaCl,0.05 $\underline{M}$ Na cacd.	(45;58)	strandedness unspecified	170
2 Poly s^{4}U	280	7	0.1 $\underline{M}$ KCl,0.05 $\underline{M}$ Na cacd.	35; ~55	no hper. at 260 mμ	232
	330	7	0.1 $\underline{M}$ KCl,0.05 $\underline{M}$ Na cacd.	34;62	t_2:hper.25%	232
	330	7	0.1 $\underline{M}$ KCl,0.01 $\underline{M}$ MgCl$_2$, 0.05 $\underline{M}$ Na cacd.	32;60.5	t_2:hper.25%	232
Poly s^{2,4}U					no complex	53b
Poly Um	max	7.8	0.1 $\underline{M}$ Na$^+$,0.01 $\underline{M}$ phosp.	68.5	hper.45%	289
2 Poly Q	265	7.0	0.1 $\underline{M}$ NaCl,0.05 $\underline{M}$ cacd.	56;81		73,200
		7.0	0.1 $\underline{M}$ NaCl,0.05 $\underline{M}$ cacd., 0.01 $\underline{M}$ Mg^{++}	65;82		73,200
	OD	7.0	0.1 $\underline{M}$ NaCl,0.05 $\underline{M}$ Na cacd.	(61;84)	strandedness unspecified	170

Polymer	Method	pH	Medium	T_m °C	Note	Ref.

Poly A Complexes (Con't)

Polymer	Method	pH	Medium	T_m °C	Note	Ref.
Poly dU	max	7.8	0.1 $\underline{M}$ Na⁺,0.01 $\underline{M}$ phosp.	54	hper.52%	289
	259	7.8	0.038 $\underline{M}$ Na⁺,0.02 $\underline{M}$ phosp.	(47)	2 → 1;hper.52%	286
	259	7.8	0.434 $\underline{M}$ Na⁺,0.02 $\underline{M}$ phosp.	(66)	2 → 1;hper.52%	286
2 Poly dU	259	7.8	0.038 $\underline{M}$ Na⁺,0.02 $\underline{M}$ phosp.	(47)	2 → 1;hper.26%	286
			0.434 $\underline{M}$ Na⁺,0.02 $\underline{M}$ phosp.	(66)	2 → 1;hper.26%	286
Poly aU		7.6	0.01 $\underline{M}$ MgCl₂,0.1 $\underline{M}$ Na cacd.		no complex	203
Poly X	260,273	7.0	0.15 $\underline{M}$ Na⁺	90.0		49
2 Poly X	250	6	0.1 $\underline{M}$ NaCl,0.05 $\underline{M}$ Na acet.	>100		176
	250	7	0.1 $\underline{M}$ NaCl,0.05 $\underline{M}$ Na cacd.	59.3;86.5		176
		7	0.1 $\underline{M}$ NaCl,0.05 $\underline{M}$ Na cacd., 0.01 $\underline{M}$ MgCl₂	62.4;77		176
	255	7.0	0.05 $\underline{M}$ Na⁺,0.025 $\underline{M}$ phosp.	84	hper.37(6)%ᴾ	57
	255	7.0	0.10 $\underline{M}$ Na⁺,0.025 $\underline{M}$ phosp.	80	hper.29(10)%ᴾ	57
	255	7.0	0.20 $\underline{M}$ Na⁺,0.025 $\underline{M}$ phosp.	74	hper.29(6)%ᴾ	57
	255	7.0	0.1 $\underline{M}$ Na⁺,0.025 $\underline{M}$ phosp.	(57;80)		262
		7.0	0.1 $\underline{M}$ NaCl,0.05 $\underline{M}$ Na cacd.	86		177

Poly m¹A Complexes

Polymer	Method	pH	Medium	T_m °C	Note	Ref.
Poly X	250	7	0.1 $\underline{M}$ NaCl,0.05 $\underline{M}$ Na cacd.	66.9		176
	250	7	0.1 $\underline{M}$ NaCl,0.05 $\underline{M}$ Na cacd., 0.01 $\underline{M}$ MgCl₂		ppt.	176

Poly m⁶ A Complexes

Polymer	Method	pH	Medium	T_m °C	Note	Ref.
Poly U	260	7.4	0.08 $\underline{M}$ NaCl,8 m$\underline{M}$ phosp.	15		71
	260	7.4	0.08 $\underline{M}$ NaCl,8 m$\underline{M}$ phosp., 0.016 $\underline{M}$ MgCl₂	17		71,180
Poly X	260	7	0.1 $\underline{M}$ NaCl,0.05 $\underline{M}$ Na cacd.	66.1		176

Polymer	Method	pH	Medium	T_m °C	Note	Ref.

Poly m⁶A Complexes (Con't)

Poly X (Con't)

Polymer	Method	pH	Medium	T_m °C	Note	Ref.
	260	7	0.1 $\underline{M}$ NaCl,0.05 $\underline{M}$ Na cacd., 0.01 $\underline{M}$ MgCl$_2$	57.7		176

Poly m⁶,⁶A Complexes

Poly U	OD	7.4	0.08 $\underline{M}$ NaCl,8 m$\underline{M}$ phosp., 0.016 $\underline{M}$ MgCl$_2$		no complex	71

Poly he⁶A Complexes

Poly X	OD	6	0.1 $\underline{M}$ NaCl,0.05 $\underline{M}$ Na acet.	>100		176
	OD	7	0.1 $\underline{M}$ NaCl,0.05 $\underline{M}$ Na cacd.	78.2		176
	260	7	0.1 $\underline{M}$ NaCl,0.05 $\underline{M}$ Na cacd., 0.01 $\underline{M}$ MgCl$_2$	70.6		176

Poly n²A Complexes

Poly U	260	8.1	0.2 $\underline{M}$ Na⁺,0.04 $\underline{M}$ pyrophosp.	86		58
	1610,1657		0.21 $\underline{M}$ Na⁺,0.04 $\underline{M}$ pyrophosp.	(79)	pD = 7.95	58

Poly n²m⁶A Complexes

Poly U	260,280; IR	7.5	0.15 $\underline{M}$ Na⁺,0.02 $\underline{M}$ phosp.	46	co.	102
	OD	7.4	VAR[Na⁺];phosp.	VAR	2 → 1	102

Poly A(ox) Complexes

Poly U	260	6.5	0.1 $\underline{M}$ NaCl,0.01 $\underline{M}$ glycylgly.	(46.5)	0%^{r};hper.31.5%	225
	260	6.5	0.1 $\underline{M}$ NaCl,0.01 $\underline{M}$ glycylgly.	(37)	7.5%^{r};hper.25.5%	225
	260	6.5	0.1 $\underline{M}$ NaCl,0.01 $\underline{M}$ glycylgly.	(33.5)	9%^{r};hper.20.8%	255

Poly A(CH$_2$O) Complexes

Poly X	260	7	0.1 $\underline{M}$ NaCl,0.05 $\underline{M}$ Na cacd.	87.5		176
	260	7	0.1 $\underline{M}$ NaCl,0.05 $\underline{M}$ Na cacd., 0.01 $\underline{M}$ MgCl$_2$	77.2		176

Poly Am Complexes

Poly U	259	7	0.04 $\underline{M}$ NaCl,0.01 $\underline{M}$ cacd.	51	M_U = 0.5; 2→1	12

Polymer	Method	pH	Medium	T_m °C	Note	Ref.
Poly Am Complexes (Con't)						
Poly U (Con't)						
	259	7	0.04 $\underline{M}$ NaCl,0.01 $\underline{M}$ cacd.	50.5	$M_U = 0.67;2\to1$	12
	259	7	0.2 $\underline{M}$ NaCl,0.01 $\underline{M}$ cacd.	64	$M_U = 0.5$ and 0.67;2→1	12
	259	7	0.5 $\underline{M}$ NaCl,0.01 $\underline{M}$ cacd.	71	$M_U = 0.5;2\to1$	12
2 Poly U	259	7	0.2 $\underline{M}$ NaCl,0.01 $\underline{M}$ cacd.	50	$M_U = 0.67;3\to2$	12
	259	7	0.04 $\underline{M}$ NaCl,0.01 $\underline{M}$ cacd.	31	$M_U = 0.67;3\to2$	12
Poly Aac Complexes						
2 Poly U	260	7.2	0.15 $\underline{M}$ NaCl,0.01 $\underline{M}$ MgCl$_2$, 0.015 $\underline{M}$ Tris	57.0	88% acetylation	125,126,127
Poly isoA Complexes						
Poly I	OD	7.0	2 m$\underline{M}$ Na$^+$,Na acet.	50.5		178
	280	7.0	0.02 $\underline{M}$ NaCl,0.01 $\underline{M}$ cacd.	86.8		177
	280	7.0	0.02 $\underline{M}$ NaCl,0.01 $\underline{M}$ cacd., 0.01 $\underline{M}$ Mg^{++}	>100		177
	OD	7.0	0.15 $\underline{M}$ Na$^+$	>100		178
	OD	7.05	0.02 $\underline{M}$ Na$^+$,Na phosp.	89.4		178
Poly U	270	7.0	0.02 $\underline{M}$ NaCl,0.01 $\underline{M}$ cacd.	86.9		177
	270	7.0	0.1 $\underline{M}$ NaCl,0.05 $\underline{M}$ cacd., 0.01 $\underline{M}$ MgCl$_2$	>100		177
		7.0	0.15 $\underline{M}$ Na$^+$	>100		178
	OD	7.1	0.02 $\underline{M}$ Na$^+$,Na phosp.	88.4		178
2 Poly X	280	6	0.1 $\underline{M}$ NaCl,0.05 $\underline{M}$ Na acet.	>100		176
	280	7	0.1 $\underline{M}$ NaCl,0.05 $\underline{M}$ Na cacd.	75.5		176
	280	7	0.1 $\underline{M}$ NaCl,0.05 $\underline{M}$ Na cacd., 0.01 $\underline{M}$ MgCl$_2$	69.5		176
Poly c^{7}A Complexes						
Poly U	270	7.0	0.0 $\underline{M}$ NaCl,0.1 $\underline{M}$ Na cacd.	32		103
	270	7.0	0.1 $\underline{M}$ NaCl,0.1 $\underline{M}$ Na cacd.	38		103
	270	7.0	0.5 $\underline{M}$ NaCl,0.1 $\underline{M}$ Na cacd.	48		103

Polymer	Method	pH	Medium	T_m °C	Note	Ref.
Poly c[7]A Complexes (Con't)						
Poly U (Con't)						
	270	7.0	1.0 $\underline{M}$ NaCl,0.1 $\underline{M}$ Na cacd.	52		103
Poly dA Complexes 2 Poly T						
	260	7	0.01 $\underline{M}$ Na$^+$	53.2	hper.38.2%	8
	260	7	0.1 $\underline{M}$ Na$^+$	69.2	hper.55.5%	8
	260	7	0.15 $\underline{M}$ Na$^+$	73.2	hper.58.5%	8
Poly dT	OD	6.30	0.025 $\underline{M}$ Na phosp.	54.9$\pm$0.2		64a
	OD	6.30	0.025 $\underline{M}$ Na phosp.,2 m$\underline{M}$ nC$_z$n	VAR	$z = 2\text{-}7^w$	64a
	260	7.0	1 m$\underline{M}$ Na cacd.	60.5		6
	260	7.0	VAR[Me^{++}];1 m$\underline{M}$ Na cacd.	VAR		6
	OD	7.0	0.05 $\underline{M}$ NaClO$_4$	61.5	hper.40-50%	271
	260	7.0	0.04 $\underline{M}$ K phosp.	63.0		8
	260	7.0	0.04 $\underline{M}$ K phosp.,8 m$\underline{M}$ MgCl$_2$	73.0		8
	260	N	SSC	71.5		8
	7		SSC	69		272
	N		1/10 SSC	51		14
	N		SSC	69.5		14
	max	7.15	0.01 $\underline{M}$ KF,1 m$\underline{M}$ Tris	50.5	z	149
	OD	7.8	0.1 $\underline{M}$ Na phosp.	68.5		37
			VAR steroidal diamines	VAR	VAR N/steroid$^\&$	64b
2 Poly dT	OD	6.30	0.025 $\underline{M}$ Na phosp.,2 m$\underline{M}$ nC$_z$n	VAR	$z = 2\text{-}5$	64a
	OD	6.30	0.025 $\underline{M}$ Na phosp.,0.01 $\underline{M}$ m$_3$nC$_z$nm$_3$	VAR	$z = 2\text{-}4$	64a
2 Poly dT	265	7.2	0.15 $\underline{M}$ NaCl,0.01 $\underline{M}$ phosp.	75	hper.50%	265
	265	7.2	0.01 $\underline{M}$ NaCl,0.01 $\underline{M}$ phosp.	51	hper.50%	265
Poly U	OD	7.0	0.05 $\underline{M}$ NaClO$_4$	41.5	hper.40-50%	271
	OD	7.8	0.1 $\underline{M}$ phosp.	45.2		37

Polymer	Method	pH	Medium	T_m ^oC	Note	Ref.

Poly dA Complexes (Con't)

Polymer	Method	pH	Medium	T_m ^oC	Note	Ref.
2 Poly U	OD	6.30	0.025 $\underline{M}$ phosp.	37.8+0.2		64a
	OD	6.30	0.025 $\underline{M}$ phosp.,0.2 $\underline{mM}$ nC_zn	VAR	z = 2-7	64a
	max	7.8	0.1 $\underline{M}$ Na$^+$,0.01 $\underline{M}$ phosp.	45.5	hper.51%	289
		(7.8)	1 $\underline{M}$ Na$^+$	77		289
2 Poly e^{5}U	260	7	0.01 $\underline{M}$ Na$^+$	31.5	b.t.;hper.35%	8
	260	7	0.1 $\underline{M}$ Na$^+$	48.2	b.t.;hper.48%	8
	260	7	0.15 $\underline{M}$ Na$^+$	52	b.t.;hper.49%	8
2 Poly Um	max	7.8	0.1 $\underline{M}$ Na$^+$,0.01 $\underline{M}$ phosp.	58.5	hper.53%	289
		(7.8)	1 $\underline{M}$ Na$^+$	88		289
Poly dU	259,283.5	7.8	0.038 $\underline{M}$ Na$^+$,0.02 $\underline{M}$ phosp.	(43)	2 → 1;hper.46% (259)	286
	259,283.5	7.8	0.434 $\underline{M}$ Na$^+$,0.02 $\underline{M}$ phosp.	(65)	2 → 1;hper.50% (259)	286
	259,283.5	7.8	1.166 $\underline{M}$ Na$^+$,0.02 $\underline{M}$ phosp.	(74)	2 → 1;hper.53% (259)	286
	259,283.5	7.8	2.04 $\underline{M}$ Na$^+$,0.02 $\underline{M}$ phosp.	(55;79)	2 → 3; 3 → 1	286
2 Poly dU	259,283.5	7.8	0.038 $\underline{M}$ Na$^+$,0.02 $\underline{M}$ phosp.	(43)	2 → 1	286
	259,283.5	7.8	0.434 $\underline{M}$ Na$^+$,0.02 $\underline{M}$ phosp.	46;67	3 → 2; 2 → 1	286
	259,283.5	7.8	1.166 $\underline{M}$ Na$^+$,0.02 $\underline{M}$ phosp.	(74)	3 → 1	286
	259,283.5	7.8	1.93 $\underline{M}$ Na$^+$,0.02 $\underline{M}$ phosp.	(79)	3 → 1	286
	max	(7.8)	1 $\underline{M}$ Na$^+$	71		289
2 Poly X	255	7.0	0.20 $\underline{M}$ Na$^+$,0.025 $\underline{M}$ phosp.	54	hper.26(0)%^{P}	57

Poly (A,C) Complexes

Polymer	Method	pH	Medium	T_m ^oC	Note	Ref.
Poly U	260	6.7	0.1 $\underline{M}$ NaCl,1 $\underline{mM}$ EDTA,phosp.; I = 0.153 $\underline{M}$	22.0	A/C = 1.36; 36%U,b.t.;slow complex formation	118

Polymer	Method	pH	Medium	T_m °C	Note	Ref.
Poly d(A-C) Complexes						
Poly (T-G)	260	N	1/10 SSC	71.5	hper.20.4%	143
		(N)	SSC	90.9		84
		N	0.2 $\underline{M}$ Na⁺	91.1		84
Poly d(G-T)	265	7.0	0.02 $\underline{M}$ Na⁺,phosp.	78.9		84
	260	7.0	1/10 SSC	(75)	co.;rev. (quick cooling); irev.(fast cooling)	283
	260	7.0	SSC	91.5	co.;rev.	283
Poly d(G-s⁴T)	260,335	N	1/10 SSC	56.3	hper.21.4%(260) and 48%(335)	143
Poly d(I-T)	260	N	1/10 SSC	41.5	hper.29.0%	143
Poly (A,G) Complexes						
Poly U	260	6.7	0.1 $\underline{M}$ NaCl,0.05 $\underline{M}$ phosp., 1 m$\underline{M}$ EDTA; I = 0.153 $\underline{M}$	42.9	A/G=1.54;51%U; 6.3ʲ;hper.25.8%	118
	260	7	SSC	(61.5)	0% G	274
	260	7	SSC	(56.5)	13% G	274
	260	7	SSC	(51.5)	25% G	274
	260	7	SSC	(43.5)	43%G;b.t.	274
Poly d(A-G) Complexes						
Poly d(C-T)	260	7.0	1/10 SSC	(67.5)	co.;rev. (quick cooling); irev.(slow cooling)	283
	260	7.0	SSC	84	co.;rev.	283
			0.02 $\underline{M}$ NaCl	89-90	traces of Mg⁺⁺(?)	26
Poly (A,br⁸G) Complexes						
2 Poly U	260	7.0	0.1 $\underline{M}$ NaCl,0.01 $\underline{M}$ cacd.	(48)	A/brG = 3.36	104

Polymer	Method	pH	Medium	T_m °C	Note	Ref.
Poly (A,br^{8}G) Complexes (Con't)						
2 Poly U (Con't)		7.0	0.1 $\underline{M}$ NaCl,0.01 $\underline{M}$ cacd., 0.01 $\underline{M}$ MgCl$_2$	(66)	A/brG = 3.36	104
Poly d(A-T) Complexes						
Poly d(A-br^{5}U)	OD	N	2 m$\underline{M}$ Na$^+$,phosp.	(~28)	s	108
Poly (A,U) Complexes						
Poly U	259	6.5	0.1 $\underline{M}$ KCl,0.01 $\underline{M}$ Na acet.	(36)	A/U = 1.63	338
Poly (A,X) Complexes						
Poly U	255	7.8	0.05 $\underline{M}$ Na$^+$,0.025 $\underline{M}$ phosp.	(48)	A/X = 6.7;2→1	262
	260	7.8	0.05 $\underline{M}$ Na$^+$,0.025 $\underline{M}$ phosp.	(38)	A/X = 1.86	262
	275	7.8	0.1 $\underline{M}$ Na,0.025 $\underline{M}$ phosp.	(41)	A/X = 0.35	262
2 Poly U	255	7.8	0.3 $\underline{M}$ Na$^+$,0.025 $\underline{M}$ phosp.	61	A/X = 6.7;3→1	262
2 Poly X	255	7.0	0.1 $\underline{M}$ Na$^+$,0.025 $\underline{M}$ phosp.	(60)	A/X = 1.86	262
	255	7.0	0.1 $\underline{M}$ Na$^+$,0.025 $\underline{M}$ phosp.	(95)	A/X = 6.7	262
Poly C Complexes						
Poly (C,G)	245	6.7	0.01 $\underline{M}$ Na cacd.		30-80$^{\circ f}$	101
Poly (C,I)	250	7.0	0.05 $\underline{M}$ NaCl,1 m$\underline{M}$ Na citr.	~45	I/C = 1.3; (I+C)/C = 1.85	101
Poly G	270	6.8	phosp.,I = 0.01 $\underline{M}$		no t. <90°	81
	280	7.0	1 m$\underline{M}$ NaCl,0.5 m$\underline{M}$ Na cacd.; 80% MeOH	~89	rev.	195
	280	7.0	1 m$\underline{M}$ NaCl,0.5 m$\underline{M}$ Na cacd.; 50% MeOH		no hper. <100°	195
	260; ORD276	7.0	1.5 m$\underline{M}$ Na$^+$;80% MeOH	90		180
	280	(N)	0.01 m$\underline{M}$ NaCl	82.5		70
	280	(N)	0.1 m$\underline{M}$ NaCl	81.5		70
	280	(N)	1 m$\underline{M}$ NaCl	77.0		70

Polymer	Method	pH	Medium	T_m °C	Note	Ref.
Poly C Complexes (Con't)						
Poly G (Con't)						
	max	7.15	0.01 $\underline{M}$ KF,1 m$\underline{M}$ Tris	90	z	149
	270	9.0	Tris; I = 0.03 $\underline{M}$		no t. <90°	81
			1 m$\underline{M}$ EDTA	~75	b.t.;hper. 25%	81
	max		e.g.	74.0		70
Poly m^{7}G	OD	7.0	1.5 m$\underline{M}$ Na$^+$;80% MeOH	69		180
Poly dG	260	6.5	0.5 m$\underline{M}$ NaCl	78		133
	260	6.5	5 m$\underline{M}$ NaCl	>100		133
		N	0.25 m$\underline{M}$ Na$_2$HPO$_4$,0.5 m$\underline{M}$ NaH$_2$PO$_4$,0.1 $\underline{M}$ EDTA	83		222
		8.4	0.66 m$\underline{M}$ NaCl,0.1 m$\underline{M}$ EDTA	64	co.;rev.	133
Poly(G,I)						
	240,250	7.0	0.1 $\underline{M}$ NaCl,0.01 $\underline{M}$ cacd.	(84)	I/G = 2	148
	OD	7.0	0.01,0.025,0.05 and 0.1 $\underline{M}$ NaCl,1 m$\underline{M}$ Na citr.	VAR	VAR G/I	148
Poly I	250	3.0	Na form.; I = 0.05 $\underline{M}$	(63)	C$^+$·I	67
	OD	VAR	VAR[Na$^+$]	VAR	C·C$^+$·I	261
	275	4.6	Na form.; I = 0.05 $\underline{M}$	53;78	g	67
	max	6	1.0 $\underline{M}$ NaCl,2 m$\underline{M}$ phosp.	(70)	b.t.	50
	250	6.15	0.1 $\underline{M}$ Na$^+$,phosp.	59.0±0.3	hper.53%[a]	64
	250	6.7	0.1 $\underline{M}$ NaCl,0.01 $\underline{M}$ Na phosp.	52	1:1 complex	71
	CAL	6.8-7.0	0.063 $\underline{M}$ Na$^+$,0.01 $\underline{M}$ citr.	54.3	d.s.hx.[h]	90
	CAL	6.8-7.0	0.104 $\underline{M}$ Na$^+$,0.01 $\underline{M}$ citr.	57.0	d.s.hx.[h]	90
	CAL	6.8-7.0	0.303 $\underline{M}$ Na$^+$,0.01 $\underline{M}$ citr.	67.8	d.s.hx.[h]	90
	CAL	6.8-7.0	0.503 $\underline{M}$ Na$^+$,0.01 $\underline{M}$ citr.	70.9	d.s.hx.[h]	90
	CAL	6.8-7.0	1.003 $\underline{M}$ Na$^+$,0.01 $\underline{M}$ citr.	74.1	d.s.hx.[h]	90
	248	6.8-7.0	VAR[Na$^+$],0.01 $\underline{M}$ citr.	VAR	co.[h]	90

Polymer	Method	pH	Medium	T_m °C	Note	Ref.

Poly C Complexes (Con't)

Poly I (Con't)

	Method	pH	Medium	T_m °C	Note	Ref.
	OD	6.9	0.1 $\underline{M}$ NaCl,0.01 $\underline{M}$ Na cacd.	58		228
	250	6.95	K phosp.; I = 0.05 $\underline{M}$	(49)		67
	248	7.0	0.01 $\underline{M}$ NaCl,0.01 $\underline{M}$ Li citr.	49.5		159
	248	7.0	0.05 $\underline{M}$ NaCl,1 m$\underline{M}$ Na citr.	57		159
	248	7.0	0.05 $\underline{M}$ NaCl,1 m$\underline{M}$ Na citr., 0.02 $\underline{M}$ MeNH$_3$Cl	(59)		159
	248	7.0	0.05 $\underline{M}$ NaCl,1 m$\underline{M}$ Na citr., 0.02 $\underline{M}$ NH$_4$Cl	(61)		159
	248	7.0	0.05 $\underline{M}$ NaCl,1 m$\underline{M}$ Na citr., 0.01 $\underline{M}$ MgCl$_2$	(77)		159
	248	7.0	0.05 $\underline{M}$ NaCl,1 m$\underline{M}$ Na citr., 5 m$\underline{M}$ sperd.	84		159
	248	7.0	0.05 $\underline{M}$ NaCl,1 m$\underline{M}$ Na citr., 0.5 m$\underline{M}$ sperm.	91		159
	248	7.0	0.05 $\underline{M}$ NaCl,1 m$\underline{M}$ Na citr., VAR[poly-L-lysine]	VAR		159
	248	7.0	0.05 $\underline{M}$ NaCl,1 m$\underline{M}$ Na citr., 0.01 $\underline{M}$ nC$_2$n	(77)		159
	248	7.0	0.05 $\underline{M}$ NaCl,1 m$\underline{M}$ Na citr., 0.01 $\underline{M}$ nC$_3$n	(79)		159
	248	7.0	0.05 $\underline{M}$ NaCl,1 m$\underline{M}$ Na citr., 0.01 $\underline{M}$ nC$_4$n	(75)		159
	248	7.0	0.05 $\underline{M}$ NaCl,1 m$\underline{M}$ Na citr., 0.01 $\underline{M}$ nC$_5$n	(70)		159
	248	7.0	0.05 $\underline{M}$ NaCl,1 m$\underline{M}$ Na citr., 0.01 $\underline{M}$ nC$_6$n	(67)		159
	248	7.0	0.05 $\underline{M}$ NaCl,1 m$\underline{M}$ Na citr., 0.01 $\underline{M}$ nC$_8$n	(61)		159
	248	7.0	0.05 $\underline{M}$ NaCl,1 m$\underline{M}$ Na citr.	57	co.;effect of poly-L-lysine on T_m	273
	245	7.0	0.1 $\underline{M}$ NaCl,0.05 $\underline{M}$ cacd.	61.3		177
	245	7.0	0.1 $\underline{M}$ NaCl,0.05 $\underline{M}$ cacd., 0.01 $\underline{M}$ Mg^{++}	69.6		177

Polymer	Method	pH	Medium	T_m $^\circ$C	Note	Ref.

Poly C Complexes (Con't)

  Poly I (Con't)

Polymer	Method	pH	Medium	T_m $^\circ$C	Note	Ref.
	268,298	7.0	0.1 M NaCl,0.01 M Na cacd. (62)		co.; rev.	214
	ORD293	7.0	0.1 M NaCl,0.01 M Na cacd. (65)		co.; rev.	214
	245	7.0	0.1 M NaCl,0.01 M phosp.	60		120
	248	7.0	0.1 M NaCl,0.01 M Li citr.	65		159
	OD	7.0	0.1 M NaCl + VAR polyamines	VAR	u	132
	250	7.0	0.125 M NaCl,0.05 M Na cacd.	53.0		181
	250	7.0	0.125 M NaCl,0.05 M Na cacd.; D_2O	53.3		181
	260	7.0	0.15 M Na$^+$	62.5		49
	OD	7.0	0.15 M Na$^+$	63		180
	248	7.0	0.15 M NaCl,6 mM Na phosp.	63.5	21.1;7.8[t]; hper.71.2%	131
	248	7.0	0.15 M NaCl,6 mM Na phosp.	62.5	16.9;4.2[t]; hper.70.7%	131
	248	7.0	0.15 M NaCl,6 mM Na phosp.	62.0	13.7;2.3[t]; hper.71.7%	131
	248	7.0	0.15 M NaCl,6 mM Na phosp.	62.5	11.1;1.2[t]; hper.67.0%	131
	248	7.0	0.15 M NaCl,6 mM Na phosp.	61.5	9.3;0.74[t]; hper.67.0%	131
	248	7.0	0.15 M NaCl,6 mM Na phosp.	61.0	8.5;0.56[t]; hper.62.0%	131
	248	7.0	0.15 M NaCl,6 mM Na phosp.	60.0	7.9;0.46[t]; hper.60.0%	131
	248	7.0	0.15 M NaCl,6 mM Na phosp.	58.5	6.7;0.28[t]; hper.58.5%	131
	248	7.0	VAR[Na$^+$]; cacd.	VAR	effect of poly-L-lys & poly-L-arg. on T_m	160
	248	7.0	VAR[Na$^+$]	VAR		159
		7	0.01 M phosp.	58		272
		7	1/10 SSC	57		272
	245	7	0.1 M NaCl,0.05 M Na cacd.	61.3		175

Polymer	Method	pH	Medium	T_m °C	Note	Ref.

Poly C Complexes (Con't)

Poly I (Con't)

Polymer	Method	pH	Medium	T_m °C	Note	Ref.
	OD	7	0.15 $\underline{M}$ Na$^+$	60.2		156
	245	7	0.15 $\underline{M}$ Na$^+$, 0.01 $\underline{M}$ Mg^{++}	69.6		175
	1657	7.2	0.399 $\underline{M}$ Na$^+$, 5 m$\underline{M}$ Na cacd., D$_2$0	69.8	h	90
		7.25	0.1 $\underline{M}$ Tris	62.0$\pm$0.3	hper.48%[a]	63
	OD	7.8	0.01 $\underline{M}$ Na$^+$,0.01 $\underline{M}$ Na phosp.	41.5	co.;hper.61%	40
	OD	7.8	0.01 $\underline{M}$ NaCl,5 m$\underline{M}$ Tris	(52)		249
		7.8	0.01 $\underline{M}$ Na$^+$,5 m$\underline{M}$ phosp.	52		288
	245	7.8	0.02 $\underline{M}$ Na$^+$,0.01 $\underline{M}$ phosp.	52.5		257
	245	7.8	0.05 $\underline{M}$ Na$^+$,0.01 $\underline{M}$ phosp.	56.5		257
	245	7.8	0.10 $\underline{M}$ Na$^+$,0.01 $\underline{M}$ phosp.	62		257
	OD	7.8	0.1 $\underline{M}$ Na$^+$,0.01 $\underline{M}$ Na phosp.	60.2	co.;hper.67%	40
	OD	7.8	0.1 $\underline{M}$ NaCl,0.01 $\underline{M}$ Tris	(61)		249
		7.8	0.1 $\underline{M}$ Na$^+$,0.5 m$\underline{M}$ EDTA,phosp.	59.6		122
	245	7.8	0.15 $\underline{M}$ Na$^+$,0.01 $\underline{M}$ phosp.	64.5		257
	245	7.8	0.20 $\underline{M}$ Na$^+$,0.01 $\underline{M}$ phosp.	66.5		257
	OD	7.8	1.0 $\underline{M}$ Na$^+$,0.01 $\underline{M}$ Na phosp.	75.3	co.;hper.66%	40
	262.5	7.8	VAR[Na$^+$];0.01 $\underline{M}$ phosp.	VAR		257
	246		0.1 $\underline{M}$ phosp.	(60)		37
			0.1 $\underline{M}$ Na$^+$;VAR % e.g.	VAR		97
	OD		VAR[Na$^+$]	VAR		287
	OD	VAR	VAR[Na$^+$]	VAR		261
			VAR steroidal diamines	VAR	VAR N/steroid[&]	64b
Poly m^{7}I	OD	7.0	0.15 $\underline{M}$ Na$^+$	39.0		180
	OD	7.0	0.15 $\underline{M}$ Na$^+$,0.01 $\underline{M}$ Mg^{++}	44.0		180
Poly dI	OD	7.8	0.01 $\underline{M}$ Na$^+$,0.01 $\underline{M}$ Na phosp.	10.1	co.;hper.40%	40
	OD	7.8	0.1 $\underline{M}$ Na$^+$,0.01 $\underline{M}$ Na phosp.	35.4	co.;hper.46%	40
	OD	7.8	1.0 $\underline{M}$ Na$^+$,0.01 $\underline{M}$ Na phosp.	52.6	co.;hper.18%	40

Polymer	Method	pH	Medium	T_m °C	Note	Ref.

Poly C Complexes (Con't)

Poly dI (Con't)

| | 246 | | 0.1 $\underline{M}$ phosp. | (35) | | 37 |
| Poly X | | 4.6;7.8 | VAR[Na$^+$] | | no complex | 57 |

Poly m^{5}C Complexes

Poly I	245	7.8	0.02 $\underline{M}$ Na$^+$,0.01 $\underline{M}$ phosp.	68.5		257
	245	7.8	0.05 $\underline{M}$ Na$^+$,0.01 $\underline{M}$ phosp.	73.5		257
	245	7.8	0.10 $\underline{M}$ Na$^+$,0.01 $\underline{M}$ phosp.	78		257
	245	7.8	0.15 $\underline{M}$ Na$^+$,0.01 $\underline{M}$ phosp.	80.5		257
	245	7.8	0.20 $\underline{M}$ Na$^+$,0.01 $\underline{M}$ phosp.	83		257
	262.5	7.8	VAR[Na$^+$];0.01 $\underline{M}$ phosp.	VAR		257
	OD	7.8	0.01 $\underline{M}$ NaCl,5 m$\underline{M}$ Tris	(71)		249
	OD	7.8	0.1 $\underline{M}$ NaCl,0.01 $\underline{M}$ Tris	(79)		249
	OD		VAR[Na$^+$]	VAR		287

Poly br^{5}C Complexes

Poly G	280	7.0	1 m$\underline{M}$ NaCl,5 m$\underline{M}$ Na cacd.; 80% MeOH		no hper.<100°	195
	280	7.0	1.5 m$\underline{M}$ Na$^+$;4% e.g.,80% MeOH	>100°	hper.starts >90°	195
Poly m^{7}G	OD	7.0	1 m$\underline{M}$ NaCl,0.5 m$\underline{M}$ Na cacd.; 80% MeOH	~90		180
Poly dG	260	6.5	0.425 m$\underline{M}$ NaH$_2$PO$_4$,0.85 m$\underline{M}$ NaH$_2$PO$_4$,0.17 m$\underline{M}$ Na$_2$EDTA	>100		205
Poly I	250	7.0	0.1 $\underline{M}$ NaCl,0.05 $\underline{M}$ cacd.	89.2		177
	250	7.0	0.1 $\underline{M}$ NaCl,0.05 $\underline{M}$ cacd., 0.01 $\underline{M}$ MgCl$_2$	~93		177
	OD	7.0	0.15 $\underline{M}$ Na$^+$	89.0		180
	OD	7.0	VAR[Na$^+$]; 5 m$\underline{M}$ cacd.	VAR		97
	245	7	0.1 $\underline{M}$ NaCl,0.05 $\underline{M}$ Na cacd.	89		175

Polymer	Method	pH	Medium	T_m $°C$	Note	Ref.
Poly br^{5}C Complexes (Con't)						
Poly I (Con't)						
	245	7	0.15 $\underline{M}$ Na$^+$,0.01 Mg^{2+}	93		175
	OD	7	0.15 $\underline{M}$ Na$^+$	89.2		156
	OD		0.1 $\underline{M}$ Na$^+$;VAR % e.g.	VAR		97
Poly m^{7}I	OD	7.0	0.15 $\underline{M}$ Na$^+$	55.5		180
		(7.0)	0.15 $\underline{M}$ Na$^+$,0.01 $\underline{M}$ Mg^{++}	58.0		180
2 Poly dI	245	6.4	0.099 $\underline{M}$ Na$^+$	(23;72)		107
	245	6.4	0.212 $\underline{M}$ Na$^+$	(56;76)		107
	245	6.4	0.561 $\underline{M}$ Na$^+$	(85)		107
Poly io^{5}C Complexes						
Poly I	245	7	0.1 $\underline{M}$ NaCl,0.05 $\underline{M}$ Na cacd.	91.2		156, 175
	245	7	0.15 $\underline{M}$ Na$^+$,0.01 $\underline{M}$ Mg^{++}	93		175
Poly C(ox) Complexes						
Poly I	249	7.4	0.1 $\underline{M}$ NaCl,0.01 $\underline{M}$ Na acet.	(49)	0%^{r};hper.36.6%	225
	249	7.4	0.1 $\underline{M}$ NaCl,0.01 $\underline{M}$ Na acet.	(44.5)	4.6%^{r};hper.34.0%	225
	249	7.4	0.1 $\underline{M}$ NaCl,0.01 $\underline{M}$ Na acet.	(41)	11.0%^{r};hper.28.4%	225
	249	7.4	0.1 $\underline{M}$ NaCl,0.01 $\underline{M}$ Na acet.	(39)	14.5%^{r};hper.15.7%	225
Poly ac^{6}C Complexes						
Poly I	OD	7.0	0.15 $\underline{M}$ Na$^+$	62.4		180
Poly Cm Complexes						
Poly I	245	7.8	0.01 $\underline{M}$ Na$^+$,5 m$\underline{M}$ phosp.	43	co.;hper.53%	288
	265	7.8	0.01 $\underline{M}$ Na$^+$,5 m$\underline{M}$ phosp.	43	co.;hper.7%	288
Poly dC Complexes						
Poly G	260	6.5	0.5 m$\underline{M}$ NaCl	78		133

Polymer	Method	pH	Medium	T_m °C	Note	Ref.
Poly dC Complexes (Con't)						
Poly G (Con't)		6.5	5 m$\underline{M}$ NaCl	>100		133
	260					
	260	8.4	0.66 m$\underline{M}$ NaCl,0.1 m$\underline{M}$ EDTA	64		133
Poly dG	260	6.5	5 m$\underline{M}$ NaH$_2$PO$_4$,2.5 m$\underline{M}$ Na$_2$HPO$_4$, 1 m$\underline{M}$ Na$_2$ EDTA;I = 0.015 $\underline{M}$	(82)	co.;rev.	205
	260	6.5	0.425 m$\underline{M}$ NaH$_2$PO$_4$,0.85 m$\underline{M}$ NaH$_2$PO$_4$,0.17 m$\underline{M}$ Na$_2$ EDTA	66		205
	260	N	2.5 m$\underline{M}$ Na$_2$HPO$_4$,5 m$\underline{M}$ NaH$_2$PO$_4$, 1 m$\underline{M}$ EDTA	81		222
	max	7.15	0.01 $\underline{M}$ KF,1 m$\underline{M}$ Tris	84.5	z	149
	IMMUN	7.4	0.014 $\underline{M}$ NaCl,1 m$\underline{M}$ Tris	(96)		244
	OD	7.5	5.34 m$\underline{M}$ Na$^+$	(75)		110
	OD	7.5	0.01 $\underline{M}$ K phosp.	78		14
Poly I		7.8	0.01 $\underline{M}$ Na$^+$,5 m$\underline{M}$ phosp.	35		288
	OD	7.8	0.01 $\underline{M}$ Na$^+$,0.01 $\underline{M}$ Na phosp.	34.8	hper.58%	40
	OD	7.8	0.1 $\underline{M}$ Na$^+$,0.01 $\underline{M}$ Na phosp.	52.3	hper.57%	40
	OD	7.8	1.0 $\underline{M}$ Na$^+$,0.01 $\underline{M}$ Na phosp.	64.3	hper.48%	40
	246		0.1 $\underline{M}$ phosp.	(52)		37
Poly dI	243	6.1	0.101 $\underline{M}$ Na$^+$	43		106
	246	6.1	0.286 $\underline{M}$ Na$^+$	51		106
	OD		VAR[Na$^+$]	VAR		106
	245	6.4	0.561 $\underline{M}$ Na$^+$	(58)		107
	245	6.4	1.01 $\underline{M}$ Na$^+$	(58)		107
		7	SSC	43		272
	246	(N)	0.1 $\underline{M}$ phosp.	(46)		37
	OD	(N)	1/10 SSC	33		14
	OD	(N)	SSC	47.5		14
	OD	7.2	0.112 $\underline{M}$ Na$^+$	(46.5)	co.; rev.	110
	OD	7.5	0.031 $\underline{M}$ Na$^+$	(36)	co.; rev.	110
	OD	6-8	VAR[Na$^+$]	VAR		110

Polymer	Method	pH	Medium	T_m °C	Note	Ref.
Poly dC Complexes (Con't)						
Poly dI (Con't)	7.8		0.01 $\underline{M}$ Na$^+$,0.01 $\underline{M}$ Na phosp.	27.5	hper.53%	40
	OD					
	OD	7.8	0.1 $\underline{M}$ Na$^+$,0.01 $\underline{M}$ Na phosp.	46.1	hper.52%	40
	OD	7.8	1.0 $\underline{M}$ Na$^+$,0.01 $\underline{M}$ Na phosp.	55	hper.53%	40
2 Poly dI	245	6.4	0.41 $\underline{M}$ Na$^+$	(38;54)		107
	245	6.4	0.561 $\underline{M}$ Na$^+$	(44;58)		107
	245	6.4	1.01 $\underline{M}$ Na$^+$	(58)		107
Poly d(m^{5}C) Complexes						
Poly I	OD	7.8	VAR[Na$^+$]; phosp.	VAR		287
Poly dI	OD	7.8	VAR[Na$^+$]; phosp.	VAR	d.s.hx.,t.s.hx.	287
Poly d(br^{5}C) Complexes						
Poly dI	243	6.1	0.101 $\underline{M}$ Na$^+$	70		106
	245	6.4	0.099 $\underline{M}$ Na$^+$	(71)		107
	245	6.4	0.212 $\underline{M}$ Na$^+$	(76)		107
	245	6.4	0.561 $\underline{M}$ Na$^+$	(85)		107
	OD	7.5	0.0105 $\underline{M}$ Na$^+$	(53.5)		110
	OD	7.5	0.212 $\underline{M}$ Na$^+$	(77.5)		110
	OD	6-8	VAR[Na$^+$]	VAR		110
Poly (C,m^{3}C) Complexes						
Poly I	250	6.7	0.1 $\underline{M}$ NaCl,0.01 $\underline{M}$ Na phosp.	~35	C/mC = 5.67; I/(C+mC) = 1.38	71
Poly (C,m^{4}C) Complexes						
Poly I	250	6.7	0.1 $\underline{M}$ NaCl,0.01 $\underline{M}$ Na phosp.	37	C/mC = 1.56; I/(C+mC) = 1	71
Poly (C,m^{4,4}C) Complexes						
Poly I	250	6.7	0.1 $\underline{M}$ NaCl,0.01 $\underline{M}$ Na phosp.	~35	C/mC = 6.7; I/(C+mC) = 1.22	71

Polymer	Method	pH	Medium	T_m °C	Note	Ref.

Poly (C,m^4,^{5}C) Complexes

Polymer	Method	pH	Medium	T_m °C	Note	Ref.
Poly I	245	7.5	0.1 $\underline{M}$ NaCl,0.01 $\underline{M}$ phosp.	(30)	C/mC = 4.9;b.t.	204
	245	7.5	0.3 $\underline{M}$ NaCl,0.01 $\underline{M}$ phosp.	(42)	C/mC = 4.9;b.t.	204
	245	7.5	0.5 $\underline{M}$ NaCl,0.01 $\underline{M}$ phosp.	(46)	C/mC = 4.9;b.t.	204
	245	7.5	0.9 $\underline{M}$ NaCl,0.01 $\underline{M}$ phosp.	(54)	C/mC = 4.9;b.t.	204
		7.5	VAR[Na$^+$]	VAR	C/mC = 4.9;b.t.	204

Poly (C,br^{5}C) Complexes

Polymer	Method	pH	Medium	T_m °C	Note	Ref.
Poly I	OD	7.0	0.1 $\underline{M}$ Na$^+$,5 m$\underline{M}$ cacd.	VAR	VAR C/brC	97
		7.0	0.3 $\underline{M}$ Na$^+$, 5 m$\underline{M}$ cacd.	VAR	VAR C/brC	97
		7.0	0.1 $\underline{M}$ Na$^+$,5 m$\underline{M}$ cacd.; 50% e.g.	VAR	VAR C/brC	97
		7.0	0.1 $\underline{M}$ Na$^+$,5 m$\underline{M}$ cacd.; 75% e.g.	VAR	VAR C/brC	97

Poly (C,ho^{4}C) Complexes

Polymer	Method	pH	Medium	T_m °C	Note	Ref.
Poly I	275	4.05	0.1 $\underline{M}$ NaCl,0.01 $\underline{M}$ acet.	(55)	C/hoC = 2.85	120
	275	4.05	0.1 $\underline{M}$ NaCl,0.01 $\underline{M}$ acet.	(65)	C/hoC = 4	120
	275	4.05	0.1 $\underline{M}$ NaCl,0.01 $\underline{M}$ acet.	(84)	100% C	120
	245	7.0	0.1 $\underline{M}$ NaCl,0.01 $\underline{M}$ phosp.	27	C/hoC = 1.5; hper.35%	120
	245	7.0	0.1 $\underline{M}$ NaCl,0.01 $\underline{M}$ phosp.	48	C/hoC = 2.85; hper.46%	120
	245	7.0	0.1 $\underline{M}$ NaCl,0.01 $\underline{M}$ phosp.	49	C/hoC = 4; hper.52%	120
	245	7.0	0.1 $\underline{M}$ NaCl,0.01 $\underline{M}$ phosp.	60	100% C;hper.61%	120
	245	7.0	0.1 $\underline{M}$ NaCl,0.01 $\underline{M}$ phosp.	49	C/hoC = 4	120
	245	7.0	0.1 $\underline{M}$ NaCl,0.01 $\underline{M}$ phosp.	48	C/hoC = 3	120
	245	7.0	0.1 $\underline{M}$ NaCl,0.01 $\underline{M}$ phosp.	27	C/hoC = 1.5	120

Poly I Complexes

Polymer	Method	pH	Medium	T_m °C	Note	Ref.
Poly X	245	7.0	0.1 $\underline{M}$ NaCl,0.05 $\underline{M}$ Na cacd.	40.2		176

Polymer	Method	pH	Medium	T_m °C	Note	Ref.
Poly I Complexes (Con't)						
Poly X (Con't)						
	245	7.0	0.1 $\underline{M}$ NaCl,0.05 $\underline{M}$ Na cacd., 0.01 $\underline{M}$ MgCl$_2$	42.5		176
	248, 260,273	7.0	0.15 $\underline{M}$ Na$^+$	40.0		49
2 Poly I Complexes						
Poly X	275	7.8	0.1 $\underline{M}$ Na$^+$	37	hper.38%^{v}	57
	275	7.8	0.2 $\underline{M}$ Na$^+$	(38)	hper.38%^{v}	57
	275	7.2	0.7 $\underline{M}$ Na$^+$	(50.5)	hper.38%^{v}	57
Poly dI Complexes						
Poly X	250,275	7.8	0.05 - 0.15 $\underline{M}$ Na$^+$		no complex	57
Poly T Complexes						
Poly X	275	7.8	0.05 $\underline{M}$ NaCl,0.01 $\underline{M}$ phosp.	54.5	hper.45%;2.0^j	57
	275	7.8	0.10 $\underline{M}$ NaCl	57	hper.45%	57
	275	7.8	0.15 $\underline{M}$ NaCl	60.5	hper.45%	57
	275	7.8	0.20 $\underline{M}$ NaCl	62	hper.42%	57
Poly U Complexes						
Poly (U,X)	275	7.8	0.1 $\underline{M}$ Na$^+$,0.025 $\underline{M}$ phosp.	41	X/U = 4;hper.57%; 8.5^j	262
	275	7.8	0.15 $\underline{M}$ Na$^+$,0.025 $\underline{M}$ phosp.	44	X/U = 4;hper.57%; 4.5^j	262
Poly X	270	6.0	0.1 $\underline{M}$ NaCl,0.05 $\underline{M}$ Na acet.	43.4		176
	260	6.0	0.15 $\underline{M}$ Na$^+$,0.01 $\underline{M}$ Mg^{++}	39.0		49
		7.0	0.1 $\underline{M}$ NaCl,0.05 $\underline{M}$ Na cacd.	49.9		176
		7.0	0.1 $\underline{M}$ NaCl,0.05 $\underline{M}$ Na cacd., 0.01 $\underline{M}$ MgCl$_2$	54.2		176
	260	7.0	0.15 $\underline{M}$ Na$^+$,0.01 $\underline{M}$ Mg^{++}	53.0		49
	260,273	7.0	0.15 $\underline{M}$ Na$^+$	50.0		49

Polymer	Method	pH	Medium	T_m °C	Note	Ref.
Poly U Complexes (Con't)						
Poly X (Con't)		7.4	0.15 $\underline{M}$ NaCl,0.075 $\underline{M}$ Tris	49.5		176
	270	7.8	0.05 $\underline{M}$ NaCl,0.01 $\underline{M}$ phosp.	43	hper.58%;1.5[j],[x]	57
	270	7.8	0.10 $\underline{M}$ NaCl,0.01 $\underline{M}$ phosp.	46.5	hper.57%;[x]	57
	270	7.8	0.15 $\underline{M}$ NaCl,0.01 $\underline{M}$ phosp.	48	hper.57%[x]	57
	270	7.8	0.20 $\underline{M}$ NaCl,0.01 $\underline{M}$ phosp.	50	hper.57%[x]	57
	275	7.8	0.1 $\underline{M}$ Na$^+$,0.025 $\underline{M}$ phosp.	46	hper.57%;1.5[j]	262
	275	7.8	0.15 $\underline{M}$ Na$^+$,0.025 $\underline{M}$ phosp.	48	hper.57%;1.5[j]	262
		7.8	0.15 $\underline{M}$ NaCl,0.075 $\underline{M}$ Tris	50.5		176
		8.4	0.15 $\underline{M}$ NaCl,0.075 $\underline{M}$ Tris	50.8		176
		9.05	0.15 $\underline{M}$ NaCl,0.075 $\underline{M}$ Tris	50.3		176
		9.50	0.15 $\underline{M}$ NaCl,0.075 $\underline{M}$ Tris	46.2		176
		10.0	0.15 $\underline{M}$ NaCl,0.075 $\underline{M}$ Tris	40.5		176
Poly e^{5}U Complexes						
Poly X	275	7.8	0.05 $\underline{M}$ NaCl,0.01 $\underline{M}$ phosp.	43.0	hper.38.5%;3.5[j]	57
	275	7.8	0.10 $\underline{M}$ NaCl,0.01 $\underline{M}$ phosp.	46	hper.37.5%	57
	275	7.8	0.15 $\underline{M}$ NaCl,0.01 $\underline{M}$ phosp.	47.5	hper.36.5%	57
	275	7.8	0.20 $\underline{M}$ NaCl,0.01 $\underline{M}$ phosp.	49	hper.39.5%	57
Poly fl^{5}U Complexes						
Poly X	275	7.8	0.05 $\underline{M}$ NaCl,0.01 $\underline{M}$ phosp.	32.5	hper.29%;3.0[j]	57
	275	7.8	0.10 $\underline{M}$ NaCl,0.01 $\underline{M}$ phosp.	35.5	hper.31%	57
	275	7.8	0.15 $\underline{M}$ NaCl,0.01 $\underline{M}$ phosp.	37	hper.31%	57
	275	7.8	0.20 $\underline{M}$ NaCl,0.01 $\underline{M}$ phosp.	37.5	hper.29%	57
Poly (U,X) Complexes						
Poly X	275	7.8	0.1 $\underline{M}$ Na$^+$,0.025 $\underline{M}$ phosp.	33	X/U = 5.25;hper.46%; 11.0[j]	262

Polymer	Method	pH	Medium	T_m °C	Note	Ref.
			3.3.4.1. Complexes of Polyvinylanalogues			
Poly v[9]Ade Complexes						
Poly U	260	7	0.05 M NaCl, 0.5 mM MgCl$_2$, 5 mM Na cacd.	(50)	irev.	191
2 Poly U	255	7.4	0.01 M NaCl, 0.01 M Tris	(47)		124
	255	7.4	0.1 M Na$^+$, 0.01 M Tris	(54)	irev.	124
	OD	7.4	VAR [Na$^+$]	VAR		124
Poly v[1]Cyt Complexes						
Poly G	260,280	10.0	0.01 M NaCl, 5 mM Na bor.; 25% p.g.	(72)	vC/G = 4;irev.[y]	190
Poly v[1]Ura Complexes						
Poly A	265	7.0	0.05 M NaCl, 1 mM MgCl$_2$, 0.01 M Na cacd.	(62)	nco.;irev.	189

FOOTNOTES FOR SECTION 3.3

[a] See the reference for T_m values of polynucleotide complexes with diamines of the general formula $R'R''R''' N(CH_2)_n NR'R''R'''$, where R', R'', and R''' may be alkyl, aryl, or hydrogen.

[b] The fractionated commercial poly A (8.2 S) exhibits a long asymmetric tail in the melting curve while the latter is reduced in the melting curve of a fractionated high-molecular-weight sample (19.3 S).

[c] The melting curve consists of a steep cooperative part at lower temperatures and a gradual noncooperative part at higher temperatures; T_m and the first hyperchromicity values are those for the cooperative transition; the second hyperchromicity value is a sum of hyperchromicities due to cooperative and noncooperative transition (between 5 and 88°C).

[d] Heating-cooling experiments.

[e] pH was not specified, but is most probably acidic (such as for protonated form of poly A).

[f] Gradual increase in optical density with increasing temperature from ca. 20 to 90°C, unless otherwise indicated.

[g] Hyperchromicity is 17% in the presence of 1% formaldehyde.

[h] Concentrations are given in moles per kilogram of water.

[i] Change in temperature (°C) required to bring the optical density transition from 10% to 90% completion.

[j] Temperature interval (°C) over which 2/3 of the total optical density change occurs.

[k] Absorbance vs. temperature profile indicated no transition.

[l] Preparations of different origins were compared.

[m] Poly d(A-T) was isolated from male sex organs of the crab _Cancer pagurus_.

[n] Temperature interval over which 90% of the total optical density change occurs.

[o] Slope of the T_m vs. -log [Na^+] relationship is -17.5°C.

[p] The first value of hyperchromicity is that after attainment of equilibrium; the one in parentheses is the value obtained immediately after mixing of components.

[q] Incomplete protonation of poly C; the second T_m value corresponds to the melting of poly C.

[r] Degree of oxydation of the adenine and cytosine moieties at N-1 and N-3, respectively.

[s] The hybrid complex is formed only at high concentrations (O.D. 30 to 100) and slow cooling in the zone of helix--coil transition. T_m is sensitive to minor variations in Na^+ concentration.

[t] The values are $S_{w,20}$ and M.W. x 10^6; the molecular weight was calculated from the expression M.W. = $1.15 \times 10^3 \times (S_{w,20})^{2.9}$.

[u] Polyamines tested were as follows: neomycin, neamine, kanamycin, spermine, spermidine, cadaverine, 1,6-diaminohexane, glucosamine, streptidine, streptomycin.

[v] No complex is formed at pH 4.6 in 0.1 - 0.2 M NaCl; mixtures of poly X and poly I other than 1:2 gave two component melting curves.

[w] See the reference for effect on T_m of aliphatic diammonium salts of the general structure $R(CH_3)_2N^+(CH_2)_zN^+(CH_3)_2R \cdot 2Br-$ (0.01 M; $z \leq 6$; R = H-, CH_3-, CH_3CH_2-, $CH_3(CH_2)_2$-, and $HO(CH_2)_4$-).

[x] No complex is formed at pH 5.2 or 5.75 in 0.05 and 0.1 M NaCl.

[y] The complex does not melt up to 100° at pH 7 in either 25% or 50% propylene glycol solutions; in 80% MeOH, the complex precipitates at increased temperature. Poly v^lCyt contained 10% of uracil residues due to hydrolysis of cytosine.

[z] See the reference for T_m values of complexes with irehdiamine A (3,20-diaminopregn-5-ene).

&
The steroidal diamines studied were as follows: $3\beta,17\beta$-diamino-
4α-androstane; 3β-dimethylamino-17β-amino-4α-androstane; and
3β-dimethylamino-17β-amino-4α-androstane; and 3β-dimethyl-
amino-17β-methylamino-4α-androstane.

REFERENCES

1. T. Ackermann: Physical states of biomolecules: Calorimetric study of
    helix-random coil transitions in solution; in "Biochemical
    Microcalorimetry"; Academic Press, New York, N. Y., 1969, p. 121-48.

2. A. J. Adler, L. Grossman, and G.D. Fasman: Polyriboadenylic and
    polydeoxyriboadenylic acids. Optical rotatory studies of pH-
    dependent conformation and their relative stability; Biochemistry, 8,
    3846-59 (1969).

3. E. O. Akinrimisi, E. Sander, and P.O.P. Ts'o: Properties of helical
    poly C; Biochemistry, 2, 340-4 (1963).

4. S. Aoyagi and Y. Inoue: Oligonucleotide studies I. Isolation of dinucleotides
    and seven trinucleotides from RNase T$_1$ digest of nucleic acids and their
    u.v. spectral characterization; J. Biol. Chem., 243, 514-20 (1968).

5. J. Applequist and V. Damle: Thermodynamics of the helix-coil equilibria in
    oligoadenylic acid from hypochromicity studies; J. Amer. Chem. Soc.,
    87, 1450-8 (1965).

6. B. Baranowska, T. Baranowski, and M. Laskowski, Sr.: Ion induced coil to helix
    transition of dialyzed poly d(A-T) from Cancer borealis; Eur. J. Biochem.,
    2, 345-53 (1968).

7. D. Barscz and D. Shugar: Influence of temperature on the stability of the acid
    and alkaline forms of poly A; Acta. Biochim. Pol., 11, 481-96 (1964).

8. D. Barszcz and D. Shugar: Complexes of poly-ribothymidylic acid with poly-adenylic
    acids and some properties of poly-deoxyriboadenylic acid: Eur. J. Biochem.,
    5, 91-100 (1968).

9. R. D. Blake and J. R. Fresco: Polynucleotides VII. Spectrophotometric study of

the kinetic of formation of the two-stranded helical complex resulting

from the interaction of polyriboadenylate and polyribouridylate;

J. Mol. Biol., 19, 145-60 (1966).

10. R. D. Blake, J. Massoulie, and J. R. Fresco: Polynucleotides VIII. A

    spectral approach to the equilibria between poly A and poly U and their

    complexes; J. Mol. Biol., 30, 291-308 (1967).

11. A. M. Bobst, F. Rottman, and P. Cerutti: Role of the ribose 2'-hydroxyl

    groups for the stabilization of the ordered structures of ribonucleic

    acid; J. Amer. Chem. Soc., 91, 4603-4 (1969).

12. A. M. Bobst, F. Rottman, and P. A. Cerutti: Effect of the methylation of the

    2'-hydroxyl group in poly A on its structure in weekly acidic and neutral

    solutions and on its capacity to form ordered complexes with poly U;

    J. Mol. Biol., 46, 221-39 (1969).

13. F. J. Bollum, E. Groeniger and M. Yoneda: Polydeoxycytidylic acid; Proc. Natl.

    Acad. Sci. U.S., 51, 853-9 (1964).

14. F. J. Bollum: Biosynthetic polydeoxynucleotides; pp.577-83 in reference 27.

15. J. Brahms: Optical activity and the conformation of polynucleotide models of

    nucleic acids; J. Mol. Biol., 11, 785-801 (1965).

16. J. Brahms, A. M. Aubertin, G. Dirheimer, and M. Grunberg-Manago: Studies of

    trinucleotide conformations. Role of guanine residues in an oligonucleotide

    chain; Biochemistry, 8, 3269-77 (1969).

17. J. Brahms, J. C. Maurizot, and A. M. Michelson: Conformational stability

    of dinucleotides in solution; J. Mol. Biol., 25, 481-95 (1967).

18. J. Brahms, J. C. Maurizot, and A. M. Michelson: Conformation and thermodynamic

    properties of oligocytidylic acids; J. Mol. Biol., 25, 465-80 (1967).

19. J. Brahms, J. C. Maurizot, and J. Pilet: Interactions contributing to the

    stability of a polynucleotide helical chain. Role of the 2'-hydroxyl

20. J. Brahms, A. M. Michelson, and K. E. Van Holde: Adenylate oligomers in single
        and double strand conformation; J. Mol. Biol., 15, 467-88 (1966).

21. F. N. Brenneman and M. F. Singer: Polynucleotide phosphorylase of
        Micrococcus lysodeikticus IV. Guanosine diphosphate as substrate
        and inhibitor; J. Biol. Chem., 239, 893-901 (1964).

22. R. L. C. Brimacombe: The secondary structure of N-methylated derivatives of
        polycytidylic acid; Biochim. Biophys. Acta., 142, 24-34, (1967).

23. R. L. C. Brimacombe and C. B. Reese: Preparation and properties of some
        methylated poly C; J. Mol. Biol., 18, 529-40 (1966).

24. D. T. Browne, J. Eisinger, and N. J. Leonard: Synthetic spectroscopic models
        related to coenzymes and base pairs II. Evidence for intramolecular
        base-base interactions in dinucleotide analogs; J. Amer. Chem. Soc.,
        90, 7302-22 (1968).

25. C. A. Bush and H. A. Sheraga: Optical activity of single-stranded poly dA and
        poly A; dependence of adenine chromophore Cotton effects on polymer
        conformation; Biopolymers, 7, 395-409 (1969).

26. C. Byrd, E. Ohtsuka, M. W. Moon, and H. G. Khorana: Synthetic deoxy-
        ribo-oligonucleotides as templates for the DNA polymers of E. coli:
        new DNA-like polymers containing repeating nucleotide sequences;
        Proc. Natl. Acad. Sci. U.S., 53, 79-86 (1956).

27. G. L. Cantoni and D. R. Davis, eds., "Procedures in Nucleic Acid Research";
        Harper and Row, New York, N. Y., 1966.

28. C. R. Cantor: Sequence dependent properties of oligonucleotides; Ph.D. Thesis,
        Univ. of California, Berkeley, 1966.

29. C. R. Cantor and W. W. Chin: Oligonucleotide interactions I. Structure of
        2:1 complexes between poly U and oligo A; Biopolymers, 6, 1745-52 (1968).

30. C. R. Cantor and I. Tinoco: Absorption and optical rotatory dispersion of seven

31. C. R. Cantor and I. Tinoco: Calculated optical properties of 64 trinucleoside
        diphosphates; Biopolymers, 5, 821-36 (1967).

32. C. R. Cantor, M. M. Warshaw, and H. Shapiro: Oligonucleotide interactions III.
        Circular dichroism studies of the conformation of deoxyoligonucleotides;
        Biopolymers, 9, 1059-77 (1970).

33. R. E. Cape and J. H. Spencer: Nucleotide clusters in DNA I. Interaction of
        purine oligonucleotides; Can. J. Biochem., 46, 1063-73 (1968).

34. G. R. Casani and F. J. Bollum: Oligodeoxythymidylate:Polydeoxyadenylate
        and oligodeoxyadenylate:polydeoxythymidylate interactions; Biochemistry,
        8, 3928-36 (1969).

35. G. Cassani and F. J. Bollum: Oligodeoxynucletide - polydeoxynucleotide
        interactions. Adenine-Thymine Base Pairs. J. Amer. Chem. Soc.,
        89, 4798-9 (1967).

36. P. Cerutti, H. T. Miles, and J. Frazier: Interaction of partially reduced
        poly U with poly A; Biochem. Biophys. Res. Commun., 22,466-72 (1966).

37. M. J. Chamberlin: Comparative properties of DNA, RNA and hybrid homopolymer
        pairs; Fed. Proc., 24, 1446-57 (1965).

38. M. J. Chamberlin: rAU copolymer; pp.513-19 in reference 27.

39. M. J. Chamberlin, R. L. Baldwin, and P. Berg: An enzymically synthesized
        RNA of alternating base sequence: physical and chemical character-
        ization; J. Mol. Biol., 7, 334-49 (1963).

40. M. J. Chamberlin and D. L. Paterson: Physical and chemical characterization
        of the ordered complexes formed between polyinosinic acid, poly-
        cytidylic acid and their deoxyribo-analogues; J. Mol. Biol., 12,
        410-28 (1965).

41. J. Clauwaert: Interactions of polynucleotides and their components IV.
        Acid denaturation of polyadenylic-polyuridylic complexes;

Z. Naturforsch. B, 23, 454-62 (1968).

42. J. Clauwaert and J. Stockx: Interactions of polynucleotides and their
    components I. Dissociation constants of the bases and their derivatives.
    Z. Naturforsch. B, 23, 25-30 (1968).

43. J. Clauwaert, J. Stockx, and L. Vandendriessche: Interaction of polynucleotides
    and their components III. Spectrophotometric study of the thermodynamic
    values involved in poly A-poly U interactions; Z Naturforsch. B, 23,
    33-8 (1968).

44. Y. Courtois, W. Guschlbauer, and P. Fromageot: Interactions spécifiques entre
    poly (C) et la guanosine eu milieu acide; Bull. Soc. Chim. Biol.,
    51, 1539 (1969).

45. R. A. Cox: Dissociation properties of RNA I. Titration of rat-liver RNA and
    model polynucleotides; Biochim. Biophys. Acta, 68, 401-10 (1963).

46. A. M. Craig and I. Isenberg: Binding of polycyclic aromatic hydrocarbons to
    poly A; Proc. Natl. Acad. Sci. U.S., 67, 1337-44, (1970).

47. M. Daune, C. A. Dekker, and H. K. Schachman: Complexes of silver ion with
    natural and synthetic polynucleotides; Biopolymers, 4, 51-76 (1966).

48. E. De Clerq, F. Eckstein, H. Sternbach, and T. C. Merigan: The antiviral
    activity of thiophosphate - substituted polyribonucleotides in vitro
    and in vivo; Virology, 42, 421-8 (1970).

49. E. De Clerg and T. C. Merigan: Requirement of a stable secondary structure
    for the antiviral activity of polynucleotides; Nature, 222, 1148-52
    (1969).

49a. M. Dourlent, J. C. Thrierr, F. Brun, and M. Leng: Formation of hairpin
     helices in uridylic acid oligomers; Biochem. Biophys. Res. Commun.,
     41, 1590-6 (1970).

50. P. Doty, H. Boedker, J. R. Fresco, R. Haselkorn, and M. Litt: Secondary
        structure in ribonucleic acids; <u>Proc. Natl. Acad. Sci. U.S.</u>, <u>45</u>, 482-99,
        (1959).

51. D. Dütting and H. G. Zachau: Spaltung einer serinspezifischen Transfer-
        ribonucleinsäure-Fraktion mit T_1-Ribonuclease; <u>Biochim. Biophys.</u>
        <u>Acta</u>, <u>91</u>, 573-83 (1964).

52. F. Eckstein and H. Gindl: Polyribonucleotides containing a phosphorothioate
        backbone; <u>Eur. J. Biochem.</u>, <u>31</u>, 558-64 (1970).

53. M. Eigen and D. Pörschke: Co-operative non-enzymatic base recognition I.
        Thermodynamics of the helix-coil transition of oligoriboadenylic acids
        at acidic pH; <u>J. Mol. Biol.</u>, <u>53</u>, 123-42 (1970).

53a. P. Faerber, W. Saenger, K. H. Scheit and D. Suck: Structural and spectral
        properties of 2,4-dithiouridine; <u>FEBS Lett.</u>, <u>10</u>, 41-5 (1970).

53b. P. Faerber and K. H. Scheit: Synthesis and properties of poly-2,4-dithiouridylic
        acid, a new analog of polyuridylic acid; <u>FEBS Lett.</u>, <u>11</u>, 11-13 (1970).

54. G. D. Fasman, C. Lindblow, and L. Grossman: The helical conformation of
        poly C: Studies on the forces involved.; <u>Biochemistry</u>, <u>3</u>, 1015-21 (1964).

55. H. Feldmann, D. Dütting, and H. G. Zachau: Serine specific transfer ribonucleic
        acids XII. Analysis of some oligonucleotide sequences and odd
        nucleotides from serine transfer ribonucleic acid; <u>Z. Physiol. Chem.</u>,
        <u>347</u>, 236-48 (1966).

56. G. Felsenfeld and H. T. Miles: The physical and chemical properties of
        nucleic acid; <u>Ann. Rev. Biochem.</u>, <u>36</u>, 407-48 (1967).

57. M. Fikus and D. Shugar: Properties of poly X and its reactions with potentially
        complementary homopolynucleotides; <u>Acta. Biochim. Pol.</u>, <u>16</u>, 55-82 (1969).

57a. J. J. Fox, L. F. Cavalieri, and N. Chang: The identification of cytidylic
        acids a and b by spectrophotometric methods; <u>J. Amer. Chem. Soc.</u>,

_75_, 4315-17 (1957).

58. J. J. Fox and D. Shugar: Spectrophotometric studies of nucleic acid

         derivatives and related compounds as a function of pH II. Natural

         and synthetic pyrimidine nucleosides; Biochim. Biophys. Acta,

         _9_, 369-84 (1952).

59. A. Franke, F. Eckstein, K-H. Scheit, and F. Cramer: Synthese von Oligo- und

         Polynucleotiden XVI. Synthese von Desoxyoligonucleotiden mit der

         Trichloräthyl-phosphateschutzgruppe; Chem. Ber., _101_, 944-53 (1968).

60. J. R. Fresco: Some investigations on the secondary and tertiary structure of

         ribonucleic acids; in H. J. Vogel, V. Bryson and J. O. Lampen; eds.,

         "Informational Macromolecules," Academic Press, New York, N.Y., 1963,

         p.121-142.

61. J. R. Fresco and E. Klemperer: Polyriboadenylic acid, a molecular analogue

         of RNA and DNA; Ann.N.Y. Acad. Sci., _81_, 730-41 (1959).

62. J. R. Fresco and J. Massoulie: Polynucleotides V. Helix-coil transition

         of poly G; J. Amer. Chem. Soc., _85_, 1352-3 (1963).

63. E. J. Gabbay: Topography of Nucleic acid helices in solutions I. Nonidentity

         of poly A-poly U and poly I-poly C helices; Biochemistry, _5_, 3036-43 (1966).

64. E. J. Gabbay: Topography of nucleic acid helices in solution II. Structure

         of the double-stranded rA-rU, rI-rC, acid rA, and the triple-stranded

         $rA-rU_2$, and $rA-rI_2$ helices; Biopolymers, _5_, 727-47 (1967).

64a. E. J. Gabbay and R. Glaser: Topography of nucleic acid helices in solutions.

         Comparative studies of the interactions of aliphatic diammonium salts

         with double- and triple-stranded deoxyribohomopolymers, hybrid homopolymers,

         and ribohomopolymers; Biochim. Biophys. Acta, _224_, 272-5 (1970).

64b. E. J. Gabbay, R. Glaser, and B. L. Gaffney: Interaction specificity of nucleic

         acids; Ann. N.Y. Acad. Sci., _171_, 810-26 (1970).

65. R. B. Gennis and C. R. Cantor: Optical properties of specific complexes
      between complementary oligribonucleotides; Biochemistry., 9, 4714-23
      (1970).

66. H. Gellert, M. N. Lipsett, and D. R. Davies: Helix formation by guanylic acid;
      Proc. Natl. Acad. Sci. U.S., 48, 2013-18 (1962).

67. G. Giannoni and A. Rich: A polynucleotide helix containing polyinosinic
      acid and protonated polycytidylic acid; Biopolymers, 2, 399-413 (1964).

68. P. T. Gilham and H. G. Khorana: Studies on polynucleotides I. A new and
      general method for the chemical synthesis of the C_5, and C_3, inter-
      nucleotide linkage. Synthesis of deoxyribodinucleotides; J. Amer. Chem.
      Soc., 80, 6212-22 (1958).

69. R. C. Grant, S. J. Harwood, and R. D. Wells: The synthesis and characterization
      of poly d(I-C).poly d(I-C); J. Amer. Chem. Soc., 90, 4474-6 (1968).

70. G. Green and H. R. Mahler: Comparative study of polyribonucleotides in aqueous
      and glycol solutions; Biochemistry, 9, 368-87 (1970).

71. B. E. Griffin, W. J. Haslam, and C. B. Reese: Synthesis and properties of
      some methylated poly C; J. Mol. Biol., 10, 353-6 (1964).

72. B. E. Griffin, A. R. Todd, and A. Rich: A synthesis of ribothymidine-5'-
      phosphophate and its conversion to polythymidylic acid; Proc. Natl. Acac.
      Sci. U.S., 44, 1123-8 (1958).

73. M. Grunberg-Manago, W. E. Cohn, F. Pochon, and A. M. Michelson: Polynucleotide
      analogues III. Polypseudouridylic acid: synthesis and some physicochemical
      and biochemical properties; Biochim. Biophys. Acta., 80, 441-7 (1964).

74. L. Grossman, S. S. Levine, and W. S. Allison: The reaction of formaldehyde
      with nucleotides and T2 bacteriophage DNA; J. Mol. Biol., 3, 47-60
      (1961).

75. N. K. Gupta, E. Ohtsuka, V. Sgaramella, H. Buchi, A. Kumar, H. Weber,
        and H. G. Khorana: Studies on polynucleotides, LXXXVIII. Enzymatic
        joining of chemically synthesized segments corresponding to the gene
        for alanine-tRNA; Proc. Natl. Acad. Sci. U.S., 60, 1338-44 (1968).

76. W. Guschlbauer: Structures des polynucléotides protonés. Un nouveau
        complexe entre l'acide polyadenylique et polycytidylique; C. R. Acad.
        Sci., Ser.D, 265, 1422-6 (1967).

77. W. Guschlbauer: Protonated polynucleotide structures I. The thermal denatura-
        tion of poly C in acid solution; Proc. Natl. Acad. Sci. U.S., 57, 1441-8
        (1967).

78. W. Guschlbauer and V. Vetterl; Protonated polynucleotide structures IV.
        Thermodynamics of the melting of the acid form of poly C; FEBS Lett.,
        4, 57-60 (1969).

79. E. Hansbury, V. N. Kerr, V. E. Mitchell, R. L. Ratliff, D. A. Smith, D. L.
        Williams, and F. M. Hayes: Synthesis of polydeoxynucleotides
        using chemically modified subunits; Biochim. Biophys. Acta, 199, 322-8
        (1970).

80. K. A. Hartman and A. Rich: The tautomeric form of helical polyribogluanylic
        acid; J. Amer. Chem. Soc., 87, 2033-9 (1965).

81. R. Haselkorn and C. E. Fox: Synthesis and properties of a complex of poly G
        and poly C; J. Mol. Biol., 13, 780-90 (1965).

82. H. Hashizume and K. Imahori: Circular dichroism and conformation of natural and
        synthetic polynucleotides. J. Biochem. Tokyo , 61, 738-49 (1967).

83. H. Hayashi and F. Egami: Fractionation and properties of guanylic acid
        polymers synthesized by ribonuclease T_1; J. Biochem. Tokyo, 53, 176-80
        (1963).

84. F. N. Hayes, E. H. Lilly, R. L. Ratliff, D. A. Smith, and D. L. Williams:
        Thermal transition in mixtures of polydeoxyribodinucleotides; Biopolymers,

9, 1103-17 (1970).

85. G. K. Helmkamp and P. O. P. Ts'o: Secondary structure of nucleic acids in
         organic solutions III. Relationship of optical properties to conforma-
         tion; Biochim. Biophys. Acta, 55, 601-8 (1962).

86. S. Hendler, E. Fürer, and P. R. Srinivasan: Synthesis and chemical properties
         of monomers and polymers containing 7-methylguanine and an investigation
         of their substrate or template properties for bacterial deoxyribonucleic
         acid or ribonucleic acid polymerases; Biochemistry, 9, 4141-51 (1970).

87. D. W. Hennage, D. M. Crothers, and D. B. Ludlum: The preparation, preservation
         and properties of high molecular weight polyadenylic acid; Biochemistry,
         8, 2298-302 (1969).

88. S. Higuchi and M. Tsuboi: Effect of polyamines on the melting of polyribouridylic
         acid complexes in a solution; Bull. Chem. Soc. Jap., 39, 1886-93 (1966).

89. S. Higuchi and M. Tsuboi: Interaction of poly-L-lysine with nucleic acids II.
         Poly (A+U), poly (A+2U), and rice dwarf virus ribonucleic acids:
         Biopolymers, 4, 837-54 (1966).

90. H. J. Hinz, W. Haar, and T. Ackermann: Experimental thermodynamics of helix-
         random coil transition III. Determination of the transition enthalpies
         of the helical complexes poly (I+C) and poly I in solution; Biopolymers,
         9, 923-36 (1970).

91. D. N. Holcomb and S. N. Timasheff: Temperature dependence of the hydrogen-
         ion equilibria in poly A; Biopolymers, 6, 513-29 (1968).

92. D. N. Holcomb and I. Tinoco: Conformation of polyriboadenylic acid: pH and
         temperature dependence; Biopolymers, 3, 121-33 (1965).

93. A. Holý: Oligonucleotidic compounds XXX. Systhesis of some 5'-O-substituted
         derivatives of guanosine 2',3'-cyclic phosphate and guanylyl-(3'-5')-
         uridine; Collect. Czech. Chem. Commun., 34, 1261-76 (1969).

94. R. B. Homer and M. Shinitzky: Poly-9-vinylacridine preparation and some
        spectral properties; <u>Macromolecules</u>, <u>1</u>, 469-72 (1968).

95. F. B. Howard, J. Frazier, M. N. Lipsett, and H. T. Miles: Infrared demonstration
        of two and three strand helix formation between poly C and guanosine
        mononucleotides; <u>Biochem. Biophys. Res. Commun.</u>, <u>17</u>, 93-102 (1964).

96. F. B. Howard, J. Frazier, and H. T. Miles: A new polynucleotide complex
        stabilized by three interbase hydrogen bonds, poly-2-aminoadenylic
        acid + polyuridylic acid; <u>J. Biol. Chem.</u>, <u>241</u>, 4293-5 (1966).

97. F. B. Howard, J. Frazier, and H. T. Miles: Interaction of poly-5-bromo-
        cytidylic acid with polyinosinic acid; <u>J. Biol. Chem.</u>, <u>244</u>, 1291-1301
        (1969).

98. F. B. Howard, J. Frazier, M. F. Singer, and H. T. Miles: Helix formation
        between polyribonucleotides and purines, purine nucleosides and nucleotides
        II; <u>J. Mol. Biol.</u>, <u>16</u>, 415-39 (1966).

99. W. M. Huang and P.O.P. Ts'o: Physicochemical basis of the recognition process
        in nucleic acid interactions; <u>J. Mol. Biol.</u>, <u>16</u>, 523-43 (1966).

100. F. Hughes: Cooperative aspects of small ion, small molecule interactions
        with nucleic acids. Potassium-polyriboadenylic acid, polyribouridylic
        acid; <u>Biophys. J.</u>, <u>10</u>, 679-99 (1970).

101. T. Ichikawa, H. Fujita, K. Matsuo, and M. Tsuboi: Helix-with-loops structure
        of polynucleotides I. Poly (C+IC) and poly (C+GU); <u>Bull. Chem. Soc. Jap.</u>,
        <u>40</u>, 2272-6 (1967).

102. K. Ikeda, J. Frazier, and H. T. Miles: Poly 2-amino-6-N-methyl-adenylic acid:
        Synthesis, characterization and interaction with poly U; <u>J. Mol. Biol.</u>,
        <u>54</u>, 59-84 (1970).

103. M. Ikehara and T. Fukui: Some physical properties of poly-7-deazaadenylic
        acid (polytubercidin phosphoric acid); <u>J. Mol. Biol.</u>, <u>38</u>,437-41 (1968).

104. M. Ikehara, I. Tazawa, and T. Fukui: Polynucleotides VII. Synthesis of
     ribonucleotides containing 8-substituted purine nucleotides by
     polynucleotide phosphorylase; Biochemistry, 8, 736-43 (1969).

105. M. Ikehara, S. Uesugi, and M. Yasumoto: A highly stacked dinucleoside
     monophosphates derived from adenine S-cyclonucleosides; J. Amer. Chem.
     Soc., 95, 4735-6 (1970).

106. R. B. Inman: Transition of DNA homopolymers; J. Mol. Biol., 9, 624-37 (1964).

107. R. B. Inman: Multistranded DNA homopolymer interactions; J. Mol. Biol.,10,
     137-146 (1964).

108. R. B. Inman and R. L. Baldwin: Formation of hybrid molecules from two alterna-
     ting DNA copolymers; J. Mol. Biol., 5, 185-200 (1962).

109. R. B. Inman and R. L. Baldwin: Helix-random coil transitions in synthetic
     deoxyribonucleic acids of alternating sequence; J. Mol. Biol., 5, 172-84
     (1962).

110. R. B. Inman and R. L. Baldwin: Helix-random coil transition in DNA homopolymer
     pairs; J. Mol. Biol., 8, 1452-69 (1964).

111. Y. Inoue, S. Aoyagi, and K. Nakanishi: Oligonucleotide studies II. Optical
     rotatory dispersion of two pairs of sequence isomers of trinucleotides.
     Tetrahedron Lett., 37, 3575-7 (1967).

112. Y. Inoue, S. Aoyagi, and K. Nakanishi: Oligonucleotide studies III.
     Optical rotatory dispersion of seven trinucleotides obtained from
     ribonuclease T_1 digests; J. Amer. Chem. Soc., 89, 5701-6 (1967).

113. Y. Inoue and K. Satoh: Oligonucleotide studies VIII. Optical rotatory disper-
     sion of five homodinucleotides; Biochem. J., 113, 843-52 (1969).

114. S. Irie, F. Egami, and Y. Inoue: Oligonucleotide studies VII. ORD of
     adenylyl-(3'-5')-4-thiouridine and guanylyl-(3'-5')-4-thiouridine;
     J. Amer. Chem. Soc., 91, 1582-4 (1969).

115. S. Irie, T. Uchida, and F. Egami: Synthesis and ribonuclease degradation of
     dinucleoside monophosphates containing a thionucleoside; Biochim.
     Biophys. Acta., 209, 289-95 (1970).

116. IUPAC - IUB Commission on Biochemical Nomenclature: Abbreviations and
     symbols for nucleic acids, polynucleotides and their constituents;
     J. Biol. Chem., 245, 5171-6 (1970); Eur. J. Biochem., 15, 203-8 (1970);
     Biochem. J., 120, 449-54(1970).

117. S. Z. Jakabhazy and S. W. Fleming: Electro-optical studies of conformation
     and interaction of polynucleotides; Biopolymers, 4, 793-813 (1966).

118. B. Janik, Unpublished results, 1970

119. C. Janion and D. Shugar: Studies on possible mechanism of hydroxylamine
     mutagenesis; Acta. Biochim. Pol., 15, 107-21 (1968).

120. C. Janion and D. Shugar: Mechanism of hydroxylamine mutagenesis: complexing
     properties of copolymers of hydroxycytidylic acid with cytidylic or
     uridylic acids; Acta. Biochim. Pol., 16, 219-33 (1969).

121. C. Janion, B. Zmudzka, and D. Shugar: The preparation of 2'-0-methylcytidine
     -5'-mono and pyrophosphate and poly-2'-0-methylcytidylic acid;
     Acta Biochim. Pol., 17, 31-40 (1970).

122. M. Karstadt and J. S. Krakow: Azotobacter vinelandii ribonuclease X. Some
     physical properties of the alternating copolymers of inosinic and
     cytidylic residues and of gnanylic and cytidylic acid residues;
     J. Biol. Chem., 245, 752-8 (1970).

123. H. Kaye: Nucleic acid analogs. The synthesis of poly-1-vinyluracil and
     poly-9-vinyladenine; J. Polym. Sci. Part B, 7, 1-5 (1969).

124. H. Kaye: Nucleic acids analogs. The interaction of poly-9-vinyladenine with
     polyuridylic acids; J. Amer. Chem. Soc., 92 5777-9 (1970).

125. D. G. Knorre and G. G. Shamovskii: Effect of acetylation of the 2'-hydroxy
     groups of polyribonucleotides on their ability to enter into complex
     formation with complementary polyribonucleotides; Mol. Biol. U.S.S.R.,
     2, 28-33 (1968).

126. D. G. Knorre and G. G. Shamovskii: Investigation of the role of 2'-hydroxy
     groups in the formation of secondary structure in polyribonucleotides:
     acetylation studies; Biochim. Biophys. Acta, 142, 555-8 (1967).

127. D. G. Knorre and G. G. Shamovskii, V. I. Sirotyuk, and L. E. Stephanovich:
     The effect of 2'-hydroxy group acetylation on the template activity
     of synthetic polynucleotides; Abh. Deutsch. Akad. Wiss. Berlin, Kl. Med.,
     1968, 519-24.

128. N. S. Kondo, H. M. Holmes, L. M. Stempel,and P.O.P. Ts'o: Influence of
     the phosphodiester linkage (3'-5', 2'-5', and 5'-5') on the conformation
     of dinucleoside monophosphate; Biochemistry., 9, 3479-98 (1970).

129. H. Krakauer: The thermodynamics of the conforational transitions of the
     poly A - poly U complexes; Ph.D. Thesis, Yale University, 1968.

130. H. Krakauer and J. M. Sturtevant: Heats of the helix-coil transitions of the
     poly A - poly U complexes; Biopolymers, 6, 491-512 (1968).

131. G. P. Lampson, A. K. Field, A. A. Tytell, M. M. Nemes, and M. R. Hilleman:
     Relationship of moleuclar size of rI_n:rC_n (Poly I:C) to induction of
     interferon and host resistance; Proc. Soc. Exp. Biol. Med., 135, 911-16
     (1970).

132. G. P. Lampson, A. A. Tytell, A. K. Field, M. M. Nemes, and M. R. Hilleman:
     Influence of polyamines on induction of interferon and resistance to
     viruses by synthetic polynucleotides; Proc. Soc. Exp. Biol., 132,
     212-18 (1969).

133. S. Lee-Huang and L. F. Cavalieri: Isolation and properties of a nucleic
     acid hybrid polymerase; Proc. Natl. Acad. Sci. U.S., 51, 1022-8 (1964).

134. S. Lee-Huang and L. F. Cavalieri: Synthesis of dA: dT; pp. 584-91 in reference 27.

135. M. Leng and G. Felsenfeld: A study of polyadenylic acid at neutral pH; J. Mol. Biol., 15, 455-66 (1966).

136. M. Leng and A. M. Michelson: Polynucléotides XI. Étude de la stabilité conformationelle de polynucléotides en: fonction de la température; Biochim. Biophys. Acta, 155, 91-7 (1968).

137. R. L. Letsinger, M. H. Caruthers, and D. M. Jerina: Reactions of nucleosides on polymer supports. Synthesis of thymidylylthymidylylthymidine; Biochemistry , 6, 1379-88 (1967).

138. R. L. Letsinger and I. S. Klaus: Investigation of a synthetic catalylic system exhibiting substrate selectivity and competitive inhibition; J. Amer. Chem. Soc., 87, 3380-6 (1965).

139. R. L. Letsinger and T. J. Savereide: Selectivity in solvolyses catalyzed by poly-(4-vinylpyridine); J. Amer. Chem. Soc., 84, 114-15, 3122-7 (1962).

140. P. A. Levene and L. W. Bass: "Nucleic acid," New York, N. Y., Chemical Catalog Co.

141. A. G. Lezius: Synthesis and characterization of a copolymer consisting of alternating deoxyadenosine and 2-thiodeoxythymidine nucleotides; Eur. J. Biochem., 14, 154-60 (1970).

142. A. G. Lezius and E. M. Gottschalk: On a reversible cooperative conformational change of a synthetic DNA depending on high ionic concentration; Z. Physiol. Chem., 351, 413-16 (1970).

143. A. G. Lezius and E. M. Gottschalk: Stabilität und Schwebedichten 4-Thiothymidin- und Inosin-substituierter alternierender Poly-deoxynucleotide Doppelstränge; Z. Physiol. Chem., 351, 119-20 (1970).

144. M. N. Lipsett: Evidence for helical structure in poly U; Proc. Natl. Acad. Sci. U.S., 46, 445-6 (1960).

145. M. N. Lipsett: The interaction of polycytidylic acid and guanine trinucleotide;
     Biochem. Biophys. Res. Commun., 11, 224-8 (1963).

146. M. N. Lipsett: Aggregation of guanine oligoribonucletides and the effect of
     mercuric salts; J. Biol. Chem., 239, 1250-60 (1964).

147. M. N. Lipsett, L. A. Heppel, and D. F. Bradley: Complex formation between
     oligonucleotides and polymers; J. Biol. Chem., 236, 857-63 (1961).

148. M. Maeda, K. Matsuo, M. Nakarnishi, and M. Tsuboi: Melting of complexes of
     polyribocytidylic acid plus copolymers of inosinic and guanylic acids
     in solution; Bull. Chem. Soc. Jap., 40, 2068-72 (1967).

149. H. R. Mahler and G. Green: Interaction of steroidal diamines with DNA and
     polynucleotides; Ann. N. Y. Acad. Sci., 171, 783-800 (1970).

149a. H. R. Mahler and B. D. Mehrotra: The interaction of nucleic acids with
      diamines; Biochim. Biophys. Acta, 68, 211-33 (1963).

150. Mann Research Laboratories, Catalog 1968.

151. J. Massoulié: Dissociation thermique de la forme acide hélicoidale de l'acide
     polyadenylique; C. R. Acad. Sci., 260, 5554-7 (1965).

152. J. Massoulié: Thermodynamique des associations de poly A et poly U en milieu
     neutre et alcalin; Eur. J. Biochem., 3, 428-38 (1968).

153. J. Massoulié: Associations de poly A et poly U en milieu acide. Phénomenes
     irréversibles; Eur. J. Biochem., 3, 439-47 (1968).

154. J. Massoulié, R. Blake, L. C. Klotz, and J. R. Fresco: Une méthode
     spectrophotométrique permettant d'étudier séparément les complexes en
     double et triple hélice formés par les acides polyadenylique et poly-
     uridylique; C. R. Acad. Sci., 259, 3104-7 (1964).

155. J. Massoulié and A. M. Michelson: Polynucléotides V. Propriétés physiques
     de dinucléotides; C. R. Acad. Sci., 259, 2923-6 (1964).

156. J. Massoulié and A. M. Michelson: Polynucléotides analogues X. Les
         reactions de déplacement entre polynucléotides; Biochim. Biophys. Acta,
         134, 22-6 (1967).

157. J. Massoulié, A. M. Michelson, and F. Pochon; Polynucleotide analogues VI.
         Physical studies on 5-substituted pyrimidine polynucleotides;
         Biochim. Biophys. Acta, 114, 16-26 (1966).

158. M. Masuda, S. Aoyagi, and Y. Inoue: Preparation of tetranucleotides,
         $A_p A_p A_p C_p$, $A_p A_p A_p G_p$ and $A_p A_p A_p U_p$; J. Biochem. Tokyo, 64, 609-11 (1968).

159. K. Matsuo and M. Tsuboi: The effects of polyamines on the melting of poly-
         riboinosinic acid plus polyribocytidylic acid complex in solution;
         Bull. Chem. Soc. Jap., 39, 347-52 (1966).

160. K. Matsuo and M. Tsuboi: Interaction of poly-L-lysine with nucleic acids ^{V}.
         Effect of Salt concentration; Biopolymers, 8, 153-6 (1969).

161. H. Matzura and F. Eckstein: A polyribonucleotide containing alternating
         $\rightarrow$P=O and $\rightarrow$P=S linkages; Eur. J. Biochem., 3, 448-52 (1968).

162. L. R. Melby and D. R. Strobach: Oligonucleotide syntheses on insoluble polymer
         supports III. Fifteen di(deoxyribonucleoside) monophosphates and several
         trinucleoside diphosphates; J. Org. Chem., 34, 427-31 (1969).

163. M. Melchers, D. Dütting, and H. G. Zachau: Enzymatische Spaltungen von Serin
         -t-RNA-Fraktionen; Biochim. Biophys. Acta, 108, 182-93 (1965).

164. F. Melchers and H. G. Zachau: Spaltung von löslicher Ribonucleosäure und
         serinspezifischen Transfer-Ribonucleinsäure-Fraktionen mit
         Pancreas-Ribonuclease; Biochim. Biophys. Acta, 91, 559-72 (1964).

165. T. C. Merigan, F. Eckstein, and E. D. A. DeClerq: Synthetic polynucleotides;
         U. S. Patent application No. 833,314, June 14, 1969.

166. A. M. Michelson: Polynucleotides I. Synthesis and properties of some poly-
         ribonucleotides; J. Chem. Soc., 1959, 1371 - 94.

167. A. M. Michelson: Polynucleotides II. Homopolymers of cytidylic and pseudo-
     uridylic acid, copolymers with repeating subunits and the stepwise
     synthesis of polyribonucleotides; J. Chem. Soc., 1959, 3655-9.

168. A. M. Michelson: Hypochromicity of oligo- and polynucleotides; Biochim.
     Biophys. Acta, 55, 841-8 (1962).

169. A. M. Michelson: "The Chemistry of Nucleosides and Nucleotides"; Academic
     Press, London, 1963, p. 446.

170. A. M. Michelson: Polynucléotides formes d'analogues de bases; Bull. Soc.
     Chim. Biol., 47, 1553-62 (1965).

171. A. M. Michelson: Oligonucleotide interactions; pp.93-106 in reference 203a.

172. A. M. Michelson, J. Dondon, and M. Grunberg-Manago: The action of polynucleo-
     tide phosphorylase on 5-halogenouridine-5'-phosphates. Biochim. Biophys.
     Acta, 55, 529-40 (1962).

173. A. M. Michelson, J. Massoulié, and W. Guschlbauer: Synthetic polynucleotides;
     in J. N. Davidson and W. Cohen, eds., Progr. Nucl. Acid Res. Mol. Biol.,
     6, 83-141 (1967).

174. A. M. Michelson and C. Monny: Polynucleotides VIII. Base stacking in poly-
     uridylic acid; Proc. Natl. Acad. Sci. U.S., 56, 1528-34 (1966).

175. A. M. Michelson and C. Monny: Polynucleotide analogues XII. Poly 5-homo-
     cytidylic acid and poly 5-iodocytidylic acid; Biochim. Biophys. Acta,
     149, 88-98 (1967).

176. A. M. Michelson and C. Monny: Polynucleotide analogues IX. Polyxanthylic
     acid; Biochim. Biophys. Acta, 129, 460-74 (1966).

177. A. M. Michelson and C. Monny: Polynucleotides X. Oligonucleotides and their
     association with polynucleotides; Biochim. Biophys. Acta, 149, 107-26
     (1967).

178. A. M. Michelson, C. Monny, R. A. Laursen, and N. J. Leonard: Polynucleotide
     analogues VIII. Poly 3-isoadenylic acid; Biochim. Biophys. Acta,

119, 258-67 (1966).

179. A. M. Michelson, C. Monny, and A. M. Kapuler: Poly 8-bromoguanylic acid;
     Biochim. Biophys. Acta, 217, 7-17 (1970).

180. A. M. Michelson and F. Pochon: Polynucleotide analogues VII. Methylation
     of polynucleotides; Biochim. Biophys. Acta, 114, 469-80 (1966).

181. H. T. Miles: A comparison of the thermal dissociation of two polynucleotide
     helices in H_2O and D_2O; Biochim. Biophys. Acta, 43, 353-5 (1960).

182. H. T. Miles: Use of inrared spectroscopy for the measurement of nucleoside
     binding to polynucleotides; in L. Grossman and K. Moldave, eds.,
     "Methods in Enzymology": Vol. XII, Part B, Academic Press, New York,
     N. Y., 1967, p. 256-67.

183. H. T. Miles and J. Frazier: A strand disproportionation reaction in a helical
     polynucleotide system; Biochem. Biophys. Res. Commun., 14, 129-36 (1964).

184. D. B. Millar and M. MacKenzie: The properties of the helix and coil forms
     of poly U and its halogenated analogues; Biochim. Biophys. Acta, 204,
     82-90 (1970).

185. T. Montenay-Garestier and C. Hélène: Interaction between cytidine and its
     cation in poly C, cytidylyl-3'-cytidine and cytidine aggregates;
     Biochemistry, 9, 2865-9 (1970).

186. R. Naylor and P. T. Gilham: Studies on some interactions and reactions of
     oligonucleotides in aqueous solution; Biochemistry, 5, 2722-8 (1966).

187. A. L. Nussbaum, G. Scheuerbrandt, and A. M. Duffied: Stepwise synthesis of
     certain deoxyribotrinucleotides; J. Amer. Chem. Soc., 86, 102-6 (1964).

188. Pharma Waldhof Catalog.

189. J. Pitha, P. M. Pitha, and P.O.P. Ts'o: Poly (1-vinyluracil): The preparation
     and interaction with adenosine derivatives; Biochim. Biophys. Acta, 204,
     39-48 (1970).

190. P. M. Pitha and A. M. Michelson: Preparation and properties of poly-
(1-vinylcytosine); Biochim. Biophys. Acta, 204, 381-8 (1970).

191. P. M. Pitha and J. Pitha: Preparation and properties of Poly-9-vinyladenine
Biopolymers, 9, 965-78 (1970).

192. P. M. Pitha and P.O.P. Ts'o: The interaction of adenosine and adenine
heptanucleoside hexaphosphate with poly U; Biochemistry, 8, 5206-16
(1969).

193. F. Pochon, J. Massoulié, and A. M. Michelson: Etude sur le "poly dAT" de crabe;
Bull. Soc. Chim. Biol., 47, 1741 (1965).

194. F. Pochon, J. Massoulié, and A. M. Michelson: Polynucléotides VII. Les DNA
satellites du thymus de veau et "Poly d(A-T)" de crabe; Biochim.
Biophys. Acta, 119, 249-57 (1966).

195. F. Pochon and A. M. Michelson: Polynucleotides VI. Interaction between poly G
and poly C; Proc. Natl. Acad. Sci. U.S., 53, 1425-30 (1965).

196. F. Pochon and A. M. Michelson: Polynucleotide analogues XI. Poly N'-
methylguanylic acid and other methylated polynucleotides; Biochim.
Biophys. Acta, 145, 321-7 (1967).

197. F. Pochon and A. M. Michelson: Polynucleotides IX. Methylation of nucleic
acids, homopolynucleotides and complexes; Biochim. Biophys. Acta, 149,
99-106 (1967).

198. F. Pochon and A. M. Michelson: Polynucleotide analogues XIV. Poly N^2-
dimethylguanylate; Biochim. Biophys. Acta, 182, 17-23 (1969).

199. F. Pochon and A. M. Michelson: Analogue de polynucléotides XV. Substitution
de poly G en C-8 par des sels de diazonium; Biochim. Biophys. Acta,
217, 225-31 (1970).

200. F. Pochon, A. M. Michelson, M. Grunberg-Manago, W. E. Cohn, and L. Dondon;
Polynucleotide analogues III. Polypseudouridylic acid: synthesis

and some physiochemical and biochemical properties; Biochim. Biophys. Acta, 80, 441-7 (1964).

201. D. Poland, J. N. Vournakis, and H. A. Sheraga: Cooperative interactions in single-strand oligomers of adenylic acid; Biopolymers, 4, 223-35 (1966).

202. V. A. Pospelov: Influence of histone upon the ability of poly A and poly U to form double complex; Mol. Biol. U.S.S.R., 4, 365-71 (1970).

203. R. G. Provenzale and J. Nagyvary: Synthesis and structure of polyribouridylic acid; Biochemistry, 9, 1744-52 (1970).

203a. B. Pullman, ed., "Molecular Associations in Biology"; Academic Press, New York, N. Y., 1968.

204. A. Rabczenko and W. Szer: Secondary structure of poly N^4,5-dimethylcytidylic acid and its copolymers with cytidylic acid; Acta Biochim. Pol., 14, 369-81 (1967).

205. C. M. Radding, J. Josse, and A. Kornberg: Enzymatic synthesis of DNA XII. A polymer of deoxyguanylate and deoxycytidylate, J. Biol. Chem., 237, 2869-76 (1962).

206. M. A. Ravitscher, P. D. Ross, and J. M. Sturtevant: The heat of the reaction between poly A and poly U; J. Amer. Chem. Soc., 85, 1915-18 (1963).

207. A. Rich and I. Tinoco: The effect of chain length upon hypochromism in nucleic acids and polynucleotides; J. Amer. Chem. Soc., 82, 6409-11 (1960).

208. E. G. Richards: On the analysis of melting curves of stacked polynucleotides; Eur. J. Biochem., 6, 88-92 (1968).

208a. E. G. Richards, C. P. Flessel, and J. R. Fresco: Polynucleotides VI. Molecular properties and conformation of polyribouridylic acid; Biopolymers, 1, 431-46 (1963).

209. E. G. Richards and H. Simpkins: A comparison of some properties of alternating poly(A-U) and the two stranded complex of poly A and poly U;

Eur. J. Biochem., 6, 93-7 (1968).

210. R. Riley, B. Maling, and M. J. Chamberlin: Physical and chemical characteriza-
      tion of two-and three-stranded adenine-thymine and adenine-uracil
      homopolymer complexes; J. Mol. Biol., 20, 359-89 (1966).

211. M. Riley, and A. V. Paul: Two- and three-stranded complexes containing homo-
      polymers poly A and poly BrU; J. Mol. Biol., 50, 439-55 (1970).

212. G. W. Rushizky and H. A. Sober: The preparation and characterization of large
      oligoribonucleotides; in J. N. Davidson and W. Cohen, eds., Progr. Nucl.
      Acid. Res. Mol. Biol., 8, 171-207 (1968).

213. C. Sander and P.O.P. Ts'o: Interaction of nucleic acids VI. Interaction of
      purine with nucleic acids. Biopolymers, 9, 765-82 (1970).

214. P. K. Sarkar and J. T. Yang: Optical activity and the conformation of
      polyinosinic acid and several other polynucleotide complexes; Biochemistry,
      4, 1238-44 (1965).

214a. P. K. Sarkar and J. T. Yang: Optical activity of 5'-guanosine monophosphate
      gel; Biochem. Biophys. Res. Commun., 20, 346-51 (1965).

215. P. K. Sarkar and J. T. Yang: ORD and conformation of polyadenylic and poly-
      uridylic acids; J. Biol. Chem., 240, 2088-93 (1965).

216. M. T. Sarocchi, Y. Courtois, and W. Guschlbauer: Protonated polynucleotide
      structures VIII. Specific complex formation between poly C and guanosine,
      or guanylic acid; Eur. J. Biochem., 14, 411-21 (1970).

217. I. E. Scheffler, E. L. Elson, and R. L. Baldwin: Helix formation by d(TA)
      oligomers II. Analysis of the helix-coil transitions of linear and cir-
      cular oligomers; J. Mol. Biol., 48, 145-164 (1970).

218. K. H. Scheit: Untersuchungen an Poly-5-hydroxymethyl Uridylsäure und Poly-5-
      methyluridylsäure; Biochim. Biophys. Acta, 134, 17-21 (1967).

219. K. H. Scheit: Polynucleotide mit Uridin und 4-Thiouridin; Biochim. Biophys.
     Acta, 145, 535-7 (1967).

220. K. H. Scheit: Enzymatic polymerization of S^4-methyl-4-thiouridine-5'-diphosphate
     by polynucleotide phosphorylase from E. coli; Biochim. Biophys. Acta,
     209, 445-54 (1970).

221. K. H. Scheit and E. Gaertner: Über die Eigenschaften von Polynucleotiden,
     welche Uridin und 4-Thiouridin enthalten; Biochim. Biophys. Acta,
     182, 10-16 (1969).

222. C. L. Schildkraut, J. Marmur, J. R. Fresco, and P. Doty: Formation and
     properties of polyribonucleotide-polydeoxyribonucleotide helical complexes;
     J. Biol. Chem., 236, PC2-4 (1961).

223. J. F. Scott and P. C. Zamecnik: Some optical properties of diadenosine-5'-
     phosphates; Proc. Natl. Acad. Sci. U.S., 64, 1308-14 (1969).

224. R. L. Scruggs and P. D. Ross: A calorimetric study of monomer-polymer complexes
     formed by poly U and some adenine derivatives; J. Mol. Biol., 47, 29-40 (1970).

225. H. Seidel: Vergleichende Untersuchungen über die Stabilitätsänderung der
     Doppelstränge poly A·poly U und poly I·poly C nach partieller N-Oxydation
     des A- beziehungsweise C-Stränges; Biochim. Biophys. Acta, 129, 414-16
     (1966).

226. D. Shugar, M. Swierkowski, M. Fikus, and D. Barszcz: Applications of model
     polynucleotides to studies on structure and function of nucleic acids;
     Abstracts 7th Internat. Congr. Biochem., Tokyo, 1967, Vol. I, Symp. 1,
     p.59-60.

227. D. Shugar and W. Szer: Secondary structure in polyribothymidylic acid
     (Poly-rT); J. Mol. Biol., 5, 580-2 (1962).

228. R. B. Sigler, D. R. Davies, and H. T. Miles: A displacement reaction between
     a polynucleotide helix and a random coil; J. Mol. Biol., 5, 509-17 (1962).

229. H. Simpkins and E. G. Richards: Titration properties of some dinucleotides;

          Biochemistry, 6, 2513-20 (1967).

230. H. Simpkins and E. G. Richards: Spectrophotometric titration studies on

          poly U; Biopolymers, 5, 551-60 (1967).

231. H. Simpkins and E. G. Richards: Preparation and properties of oligo-uridylic

          acid; J. Mol. Biol., 29, 349-56 (1967).

232. J. Simuth, K. H. Scheit, and E. M. Godschaelk: The enzmatic synthesis of

          poly 4-thiouridylic acid by polynucleotide phophorylase from E. coli;

          Biochim. Biophys. Acta, 204, 371-80 (1970).

233. M. F. Singer, L. A. Heppel, G. W. Rushizky, and H. A. Sober: Spectral properties

          of adenine oligoribonucleotides; Biochim. Biophys. Acta, 61, 474-7

          (1962).

234. J. Smrt: Homopolymers of N^4-hydroxy- and N^4-methoxycytidylic acid and their inter-

          action with polyadenylic acid; Coll. Czech. Chem. Commun., 35, 2314-23

          (1970).

235. H. A. Sober, ed.: "Handbook of Biochemistry. Selected Data for Molecular

          Biology"; The Chemical Rubber Co., Cleveland, Ohio, 1968.

236. M. Staehelin: Studies on nucleotide sequences in ribonucleic acids II.

          Spectroscopic properties of oligoribonucleotides; Biochim. Biophys.

          Acta, 49, 20-6 (1961).

237. W. M. Stanley: Physical studies on the ribosomes and ribosomal ribonucleic

          acid of Escherichia coli; Ph.D. Thesis, University of Wisconsin, 1963.

238. R. Steiner: Observations upon copolymers of adenylic and uridylic acids;

          Ann. N. Y. Acad. Sci., 81, 742-4 (1959).

239. R. F. Steiner: Copolymers of inosinic acid with cytidylic and uridylic

          acids; J. Biol. Chem., 236, 3037-42 (1961).

240. R. F. Steiner and R. Beers: Polynucleotides II. Physical properties of solutions
        of some polynucleotides; Biochim. Biophys. Acta, 26, 336-48 (1957).

241. R. F. Steiner and R. F. Beers: Polynucleotides VII. Interaction of poly A
        and poly U; Biochim. Biophys. Acta, 33, 470-81 (1959).

242. R. Steiner and R. Beers: "Polynucleotides"; Elsevier, New York, N.Y., 1961.

243. C. L. Stevens and G. Felsenfeld: The conversion of two-stranded poly (A+U) to
        three-stranded poly (A+2U) and poly A by heat; Biopolymers, 2, 293-314
        (1964).

244. B. D. Stollar: Double-helical polynucleotides: immunochemical recognition
        of differing conformations; Science, 169, 609-11 (1970).

245. P. A. Straat and P.O.P. Ts'o: Ribonucleic acid polymerase from
        Micrococcus luteus (M. lysodeikticus) IV. Effect of rifampicin on the
        homopolymer directed reaction; Biochemistry, 9, 926-31 (1970).

246. M. Swierkowski and D. Shugar: A new thymine base analogue, 5-ethyluracil:
        5-ethyluridine-5'-pyrophosphate and poly-5-ethyluridylic acid; Acta.
        Biochim Pol., 16, 763-77 (1969).

247. M. Swierkowski and D. Shugar: Poly 5-ethyluridylic acid, a polyuridylic acid
        analogue; J. Mol. Biol., 47, 57-68 (1970).

248. M. Swierkowski, W. Szer, and D. Shugar: Some properties of polyribothymidylic
        acid, copolymers of uridylic and ribothymidylic acids, and their 1:1
        complexes with polyadenylic acid; Biochem. Z., 342, 429-36 (1965).

249. W. Szer: Secondary structure of poly-5-methylcytidylic acid; Biochem.
        Biophys. Res. Commun., 20, 182-6 (1965).

250. W. Szer: Effect of di- and polyamines on the thermal transition of synthetic
        polynucleotides; Biochem. Biophys. Res. Commun., 22, 559-64 (1966).

251. W. Szer: Interaction of Poly r-T with metal ions and aliphatic amines.

Acta Biochim. Pol., 13, 251-66 (1966).

252. W. Szer: Ordered state of poly U above room temperature; J. Mol. Biol.,
         16, 585-7 (1966).

253. W. Szer and D. Shugar: The preparation and properties of high molecular weight
         polymers of N-methyluridylic acid; Acta. Biochim. Pol., 8, 235-48 (1961).

254. W. Szer and D. Shugar: Synthesis and physico-chemical and enzymatic properties
         of 5-bromo derivatives of uridine phosphates and their polymers;
         Acta. Biochim. Pol., 8, 363-74 (1961).

255. W. Szer and D. Shugar: Secondary structure in poly rT; J. Mol. Biol., 5, 580-2
         (1962).

256. W. Szer and D. Shugar: Preparation of poly-5-fluorouridylic acid and the pro-
         perties of halogenated poly-uridylic acids and their complexes with poly-
         adenylic acid; Acta. Biochim. Pol., 10, 219-31 (1963).

257. W. Szer and D. Shugar: The structure of poly-5-methylcytidylic acid and its
         twin-structured complexes with polyinosinic acid; J. Mol. Biol.,
         17, 174-87 (1966).

258. W. Szer, M. Swierkowski, and D. Shugar: Secondary structure of poly U and
         poly rT, their N-methylated analogues and their 1:1 complexes with
         poly A; Acta. Biochim. Pol., 10, 87-105 (1963).

259. I. Tazawa, S. Tazawa, L. M. Stempel, and P.O.P. Ts'o: L-adenylyl-(3'-5')-
         L-adenosine and L-adenylyl-(2'-5')-L-adenosine; Biochemistry, 9, 3499-
         3514 (1970).

260. D. Thiele and W. Guschlbauer: Protonated polynucleotide structures IV.
         Evidence for a three-stranded complex between poly(I) and poly C;
         FEBS Lett., 1, 173-5 (1968).

261. D. Thiele and W. Guschlbauer: Polynucléotides protonés VII. Transition
         thermiques entre differents complexes de l'acide polypolyinosinique

et de l'acide polycytidylique en milieu acide; <u>Biopolymers</u>, <u>8</u>, 361-78 (1969).

262. M. Tichy and M. Fikus: Investigation on the structure of xanthine-uracil and xanthine-adenine copolymers and their complexes with homopolynucleo- tides; <u>Acta Biochim. Pol.</u>, <u>17</u>, 53-71 (1970).

263. I. Tinoco, R. C. Davis, and S. R. Jaskunas: Base-base interactions in nucleic acids; pp. 77-92 in reference 203a.

264. J. N. Toal, G. W. Rushizky, A. W. Pratt, and H. A. Sober: Computer-assisted characterization of oligoribonucleotides by their ultraviolet absorption; <u>Anal. Biochem.</u>, <u>23</u>, 60-71 (1968).

265. Z. Tramer, K. L. Wierzchowski, and D. Shugar: Influence of polynucleotide secondary structure on thymine photodimerization; <u>Acta. Biochim. Pol.</u>, <u>16</u>, 83-107 (1969).

266. F. Travers, A. M. Michelson, and P. Dondon; Low-temperature studies of polynucleotides; <u>Biochim. Biophys. Acta</u>, <u>199</u>, 29-35 (1970).

267. A. R. Trim and J. E. Parker: Preparation, purification and analyses of thirteen alkali-stable dinucleotides from yeast ribonucleic acid; <u>Biochem. J.</u>, <u>116</u>, 589-98 (1970).

268. P.O.P. Ts'o: The physiochemical basis of interactions of nucleic acids; pp. 39-75 in reference 203a.

269. P.O.P. Ts'o, G. K. Helmkamp, and C. Sander: Secondary structures of nucleo- tides II. Optical properties of nucleotides and nucleic acids; <u>Biochim. Biophys. Acta</u>, <u>55</u>, 584-600 (1962).

270. P.O.P. Ts'o, G. K. Helmkamp, and C. Sander: Interaction of nucleosides and related compounds with nucleic acids as indicated by the change of helix-coil transition temperature; <u>Proc. Natl. Acad. Sci. U.S.</u>, <u>48</u>, 686-98 (1962).

271. P.O.P. Ts'o, S. A. Rapaport, and F. J. Bollum: A comprehensive study of
         polydeoxyribonucleotides and polyribonucleotides by ORD; Biochemistry,
         5, 4153-70 (1966).

272. M. Tsuboi: Helical complexes of poly-L-lysine and nucleic acids; in
         G. N. Ramachandran, ed., "Conformation of Biopolymers"; Vol. 2,
         Academic Press, New York, 1967, p. 689-702.

273. M. Tsuboi, K. Matsuo, and P.O.P. Ts'o: Interaction of poly-L-lysine and nucleic
         acids; J. Mol. Biol., 15, 256-67 (1966).

274. D. Uhlenbeck, R. Harrison, and P. Doty: Some effects of noncomplementary
         bases on the stability of helical complexes of polyribonucleotides;
         pp. 107-14 in reference 203a.

275. K. E. Van Holde, J. Brahms, and A. M. Michelson: Base interactions of
         nucleotide polymers in aqueous solutions; J. Mol. Biol., 12, 726-39
         (1965).

276. T. V. Venkstern and A. A. Baev: "Absorption Spectra of Minor Bases, Their
         Nucleosides, Nucleotides and Selected Oligoribonucleotides"; Plenum
         Press Data Division, New York, N. Y., 1965.

277. D. C. Ward, A. Cerami, and E. Reich: Biochemical studies of the nucleoside
         analogue, formycin; J. Biol. Chem., 244, 3243-50 (1969).

278. R. C. Warner: Studies of polynucleotides synthesized by polynucleotide
         phosphorylase III. Interactions and ultraviolet absorption; J. Biol. Chem.,
         229, 711-24 (1957).

279. R. C. Warner: The stability of polynucleotide complexes; pp. 111-20 in
         reference 60.

280. M. M. Warshaw and I. Tinoco: Absorption and optical rotatory dispersion of
         six dinucleotide phosphates; J. Mol. Biol., 13, 54-64 (1965).

281. M. M. Warshaw and I. Tinoco: Optical properties of sixteen dinucleoside
     phosphates; J. Mol. Biol., 20, 29-38 (1966).

282. W. J. Wechter: Nucleic Acids I. The synthesis of nucleotides and dinucleoside
     phosphates containing ara-cytidine; J. Med. Chem., 10, 762-73 (1967).

283. R. D. Wells, E. Ohtsuka, and H. G. Khorana: Studies on polynucleotides L.
     Synthetic deoxyribopolynucleotides as templates for DNA polymerase
     of E. coli: a new double-stranded DNA-like polymer containing repeating
     dinucleotide sequences; J. Mol. Biol., 14, 221-40 (1965).

283a. J. T. Yang and T. Samejima: Optical rotatory dispersion and circular dichroism
     of nucleic acids; in J. N. Davidson and W. Cohen, eds., Progr. Nucl.
     Acid Res. Mol. Biol., 9, 224-300 (1969).

284. H. G. Zachau, D. Dütting, and H. Feldman: Serine specific transfer ribonucleic
     acids XI. The structures of two serine transfer ribonucleic acids;
     Z. Physiol. Chem., 347, 212-35 (1966).

285. Ch. Zimmer and W. Szer: Interaction of copper (II) ions with poly C and
     methylated derivatives; Acta. Biochim. Pol., 15, 339-54 (1968).

286. B. Zumdzka, F. J. Bollum, and D. Shugar: Poly dU and its complexes with
     poly A and poly dA; J. Mol. Biol., 46, 169-84 (1969).

287. B. Zmudzka, F. J. Bollum, and D. Shugar: Poly-5-methyldeoxycytidylic acid and
     some alkylamino analogs; Biochemistry, 8, 3049-59 (1969).

288. B. Zmudzka, C. Janion, and D. Shugar: Poly 2'-O-methylcytidylic acid and
     the role of the 2'-hydroxyl in polynucleotide structure; Biochem.
     Biophys. Res. Commun., 37, 895-901 (1969).

289. B. Zmudzka and D. Shugar: Role of the 2'-hydroxyl in polynucleotide conforma-
     tion. Poly 2'-O-methyluridylic acid; FEBS Lett., 8, 52-4 (1970).